The Illustrated Guide to Audio Formats

Written and illustrated by
Ashley Blewer

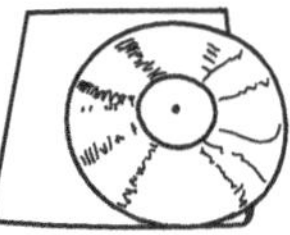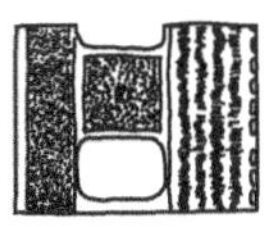

The Illustrated Guide to Audio Formats

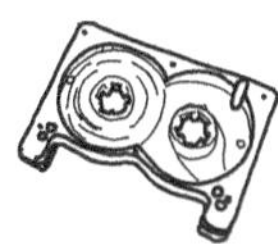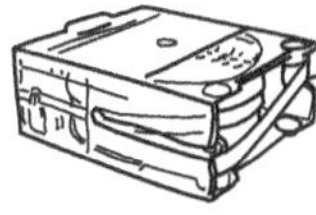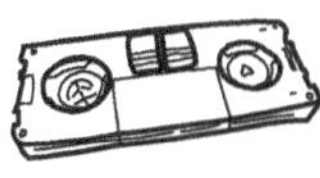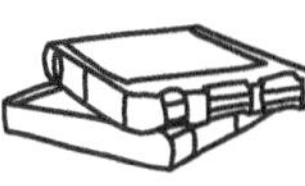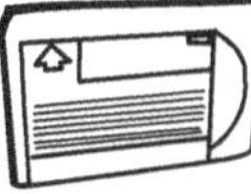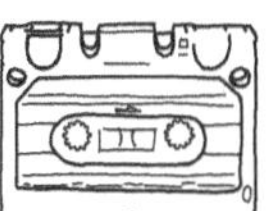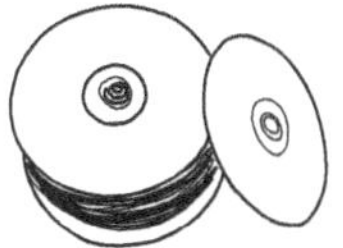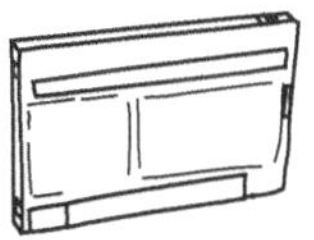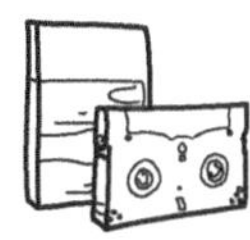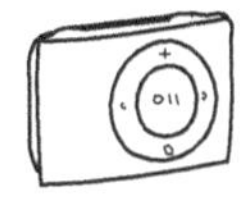

Written and illustrated by

Ashley Blewer

For Rory

Table of Contents

Introduction

This book provides an introductory overview to thirty-six
historically significant audio formats, starting with the
Phonograph and ending with digital audio players.

Each format page includes the following details:

AKA: Any other notable names that this format was "also
known as"

Format: Whether the format was primarily analog or digital

Developed by: The primary developer, manufacturer, or
patent holder(s) for the format

Era: The approximate era of production (acknowledging that
end dates are particularly fuzzy, as some formats continued
to be used long past their production expiration date)

Capacity: The maximum duration that the format was able to
hold

Size: The physical dimension(s) most common for the format,
measured in inches (for reel diameters) or centimeters (for
cartridges or other rectangular objects)

Fun facts: Three informational sentences about the format

I hope you enjoy this illustrated guide and I encourage you
to explore each of the formats more thoroughly on your own!

Phonautograph

Fun fact
Recordings from this format
were not able to be
played back until
researchers were able
to process them in 2008

Fun fact
The design of this
recording device was meant
to mimic the human ear drum

Tinfoil Phonograph

Capacity
Under a minute

Size
13 × 38 cm

Fun fact
This format was invented by Edison while working on telegram technology

Developed by
Thomas Edison

Era
1877–1900s

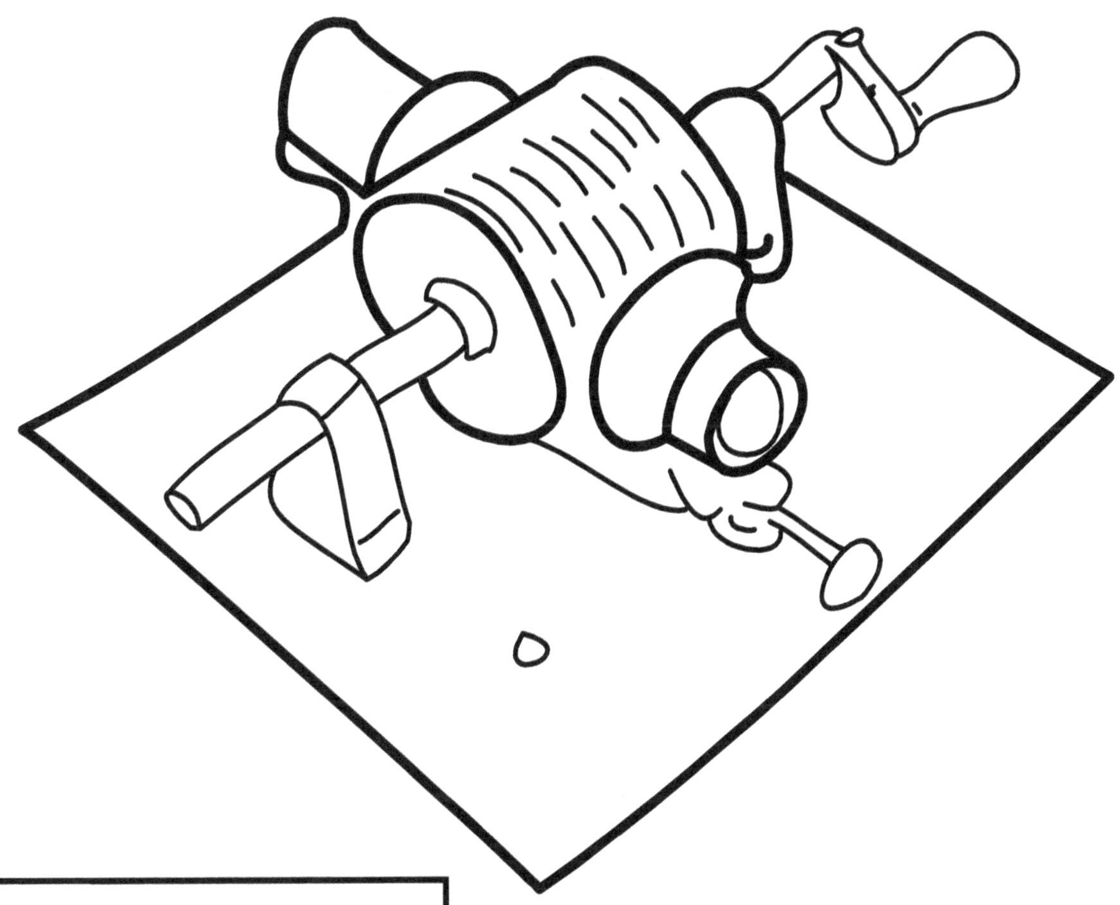

Fun fact
While this format was tinfoil
wrapped around a metal roll,
a "phonographic cylinder"
refers to its wax-coated
sibling and a "phonograph
record" refers to disc-shaped
records played by a gramophone

Fun fact
This format could be recorded
and played back by turning
a hand-crank

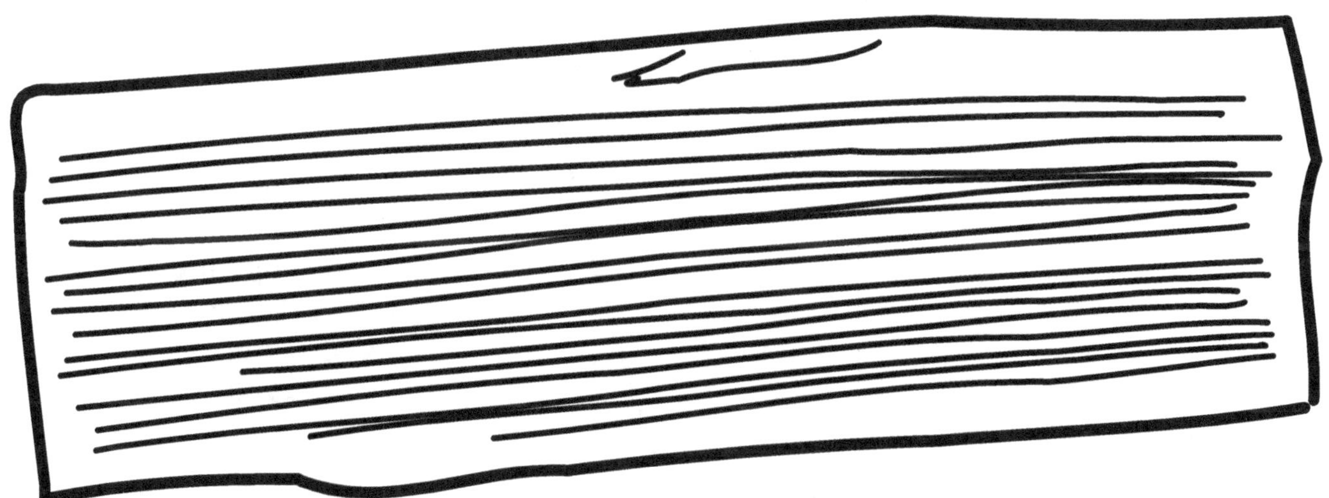

Gramophone disc

Fun fact
Earlier discs in
this format were
made from a
shellac (a resin
secreted by Iac
insects) combined
with other materials

Fun fact
Early recordings differed
in speeds, ranging
from 60-130
revolutions per minute
(rpm), but 78 rpm became
the industry-wide
standard by the
mid-1920s until the
introduction of the
microgroove disc
in the 1940s

Phonographic Cylinder

Fun fact
This format had several different material compositions: brown wax (metal soap), molded wax (metal soap), and Blue Amberol (celluloid)

Fun fact
Early brown wax cylinders would commonly wear out after they were played only a few dozen times

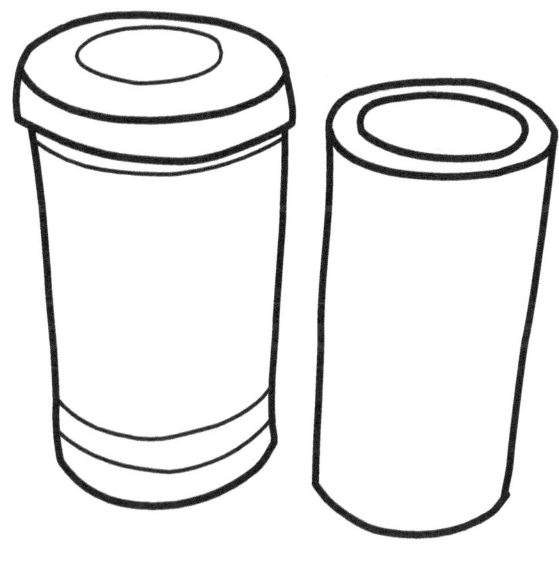

Fun fact
While this format was typically 2.25" in diameter and around 4.25" long, they could be smaller (1.33" in diameter) or larger (5" in diameter)

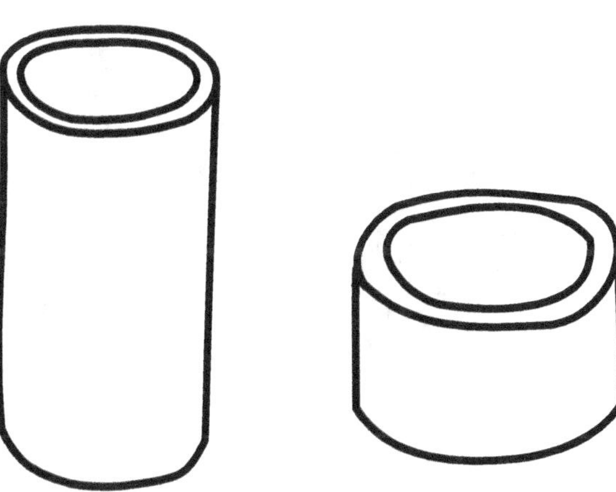

Wire Recording

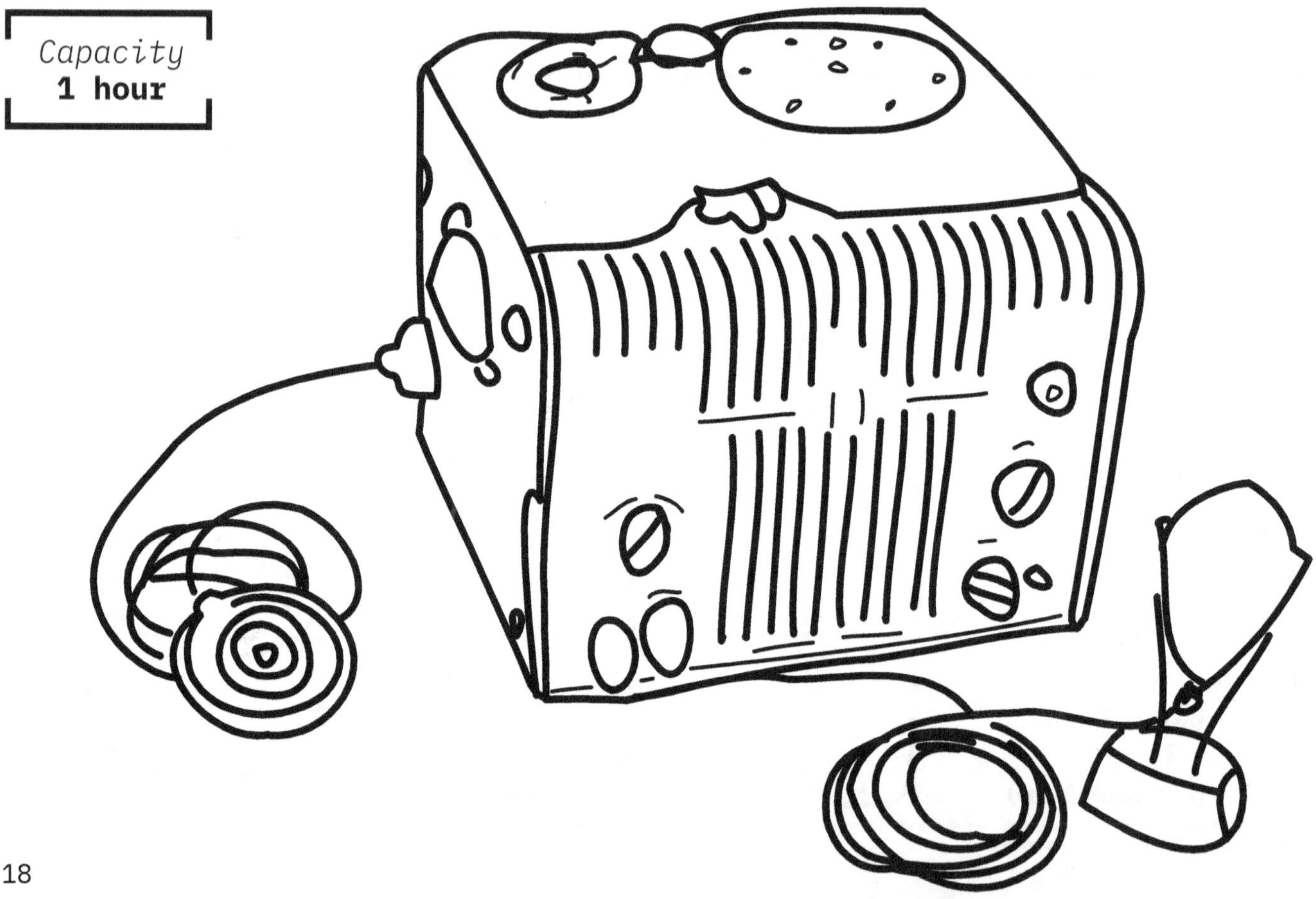

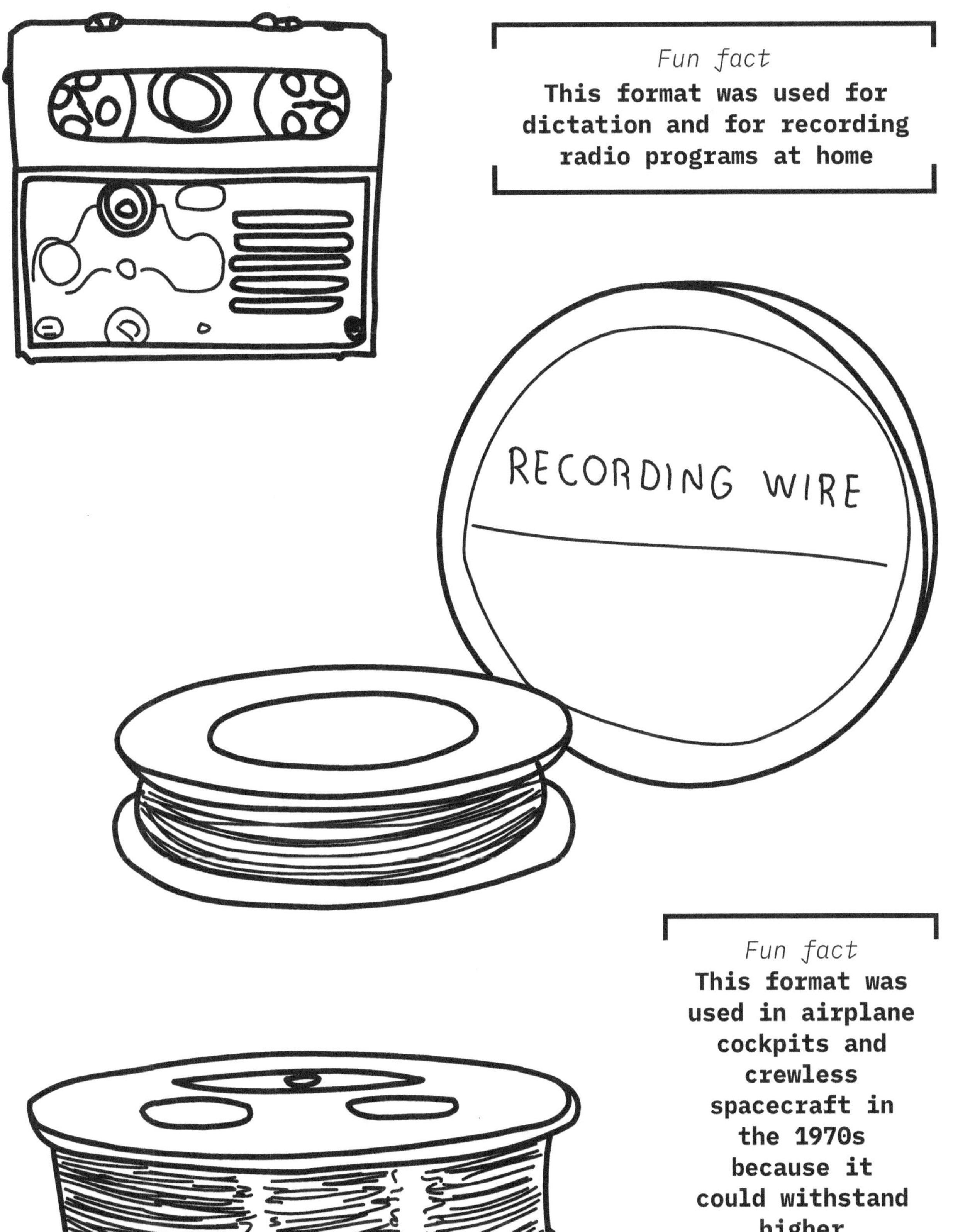

RECORDING WIRE

Fun fact
This format was used for dictation and for recording radio programs at home

Fun fact
This format was used in airplane cockpits and crewless spacecraft in the 1970s because it could withstand higher temperatures than magnetic tape

Sound on film

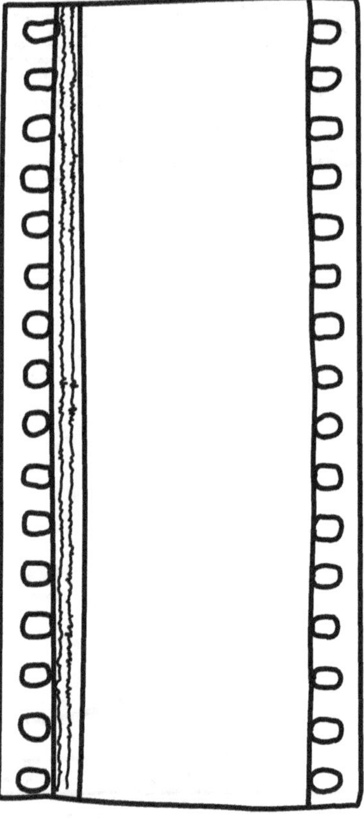

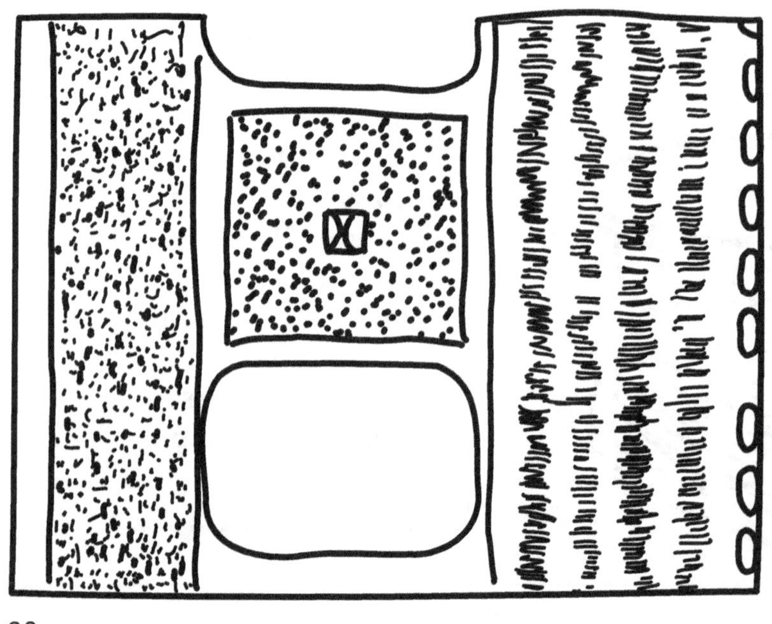

Fun fact
This format includes any process where audio is transformed into a visual representation and printed onto film, either alongside or separate from moving images

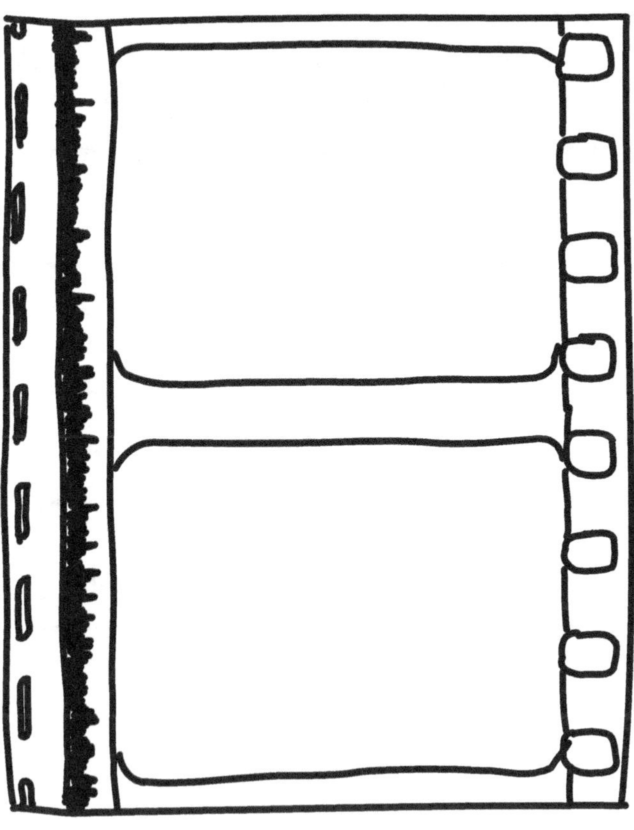

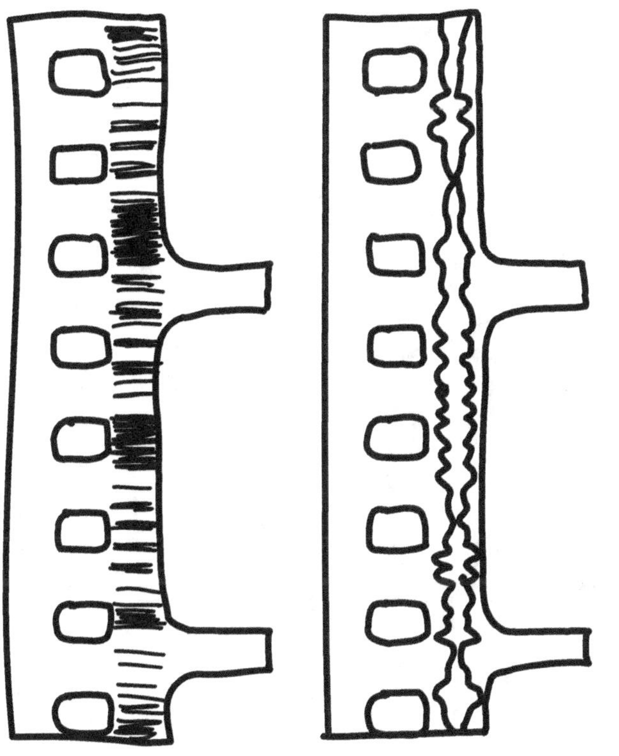

Fun fact
This process can be either an analog or digital audio track, and the signal can be recorded optically or magnetically

Fun fact
Prior to this invention, the film's soundtrack could be played on a separate phonograph record or performed live

Open reel tape

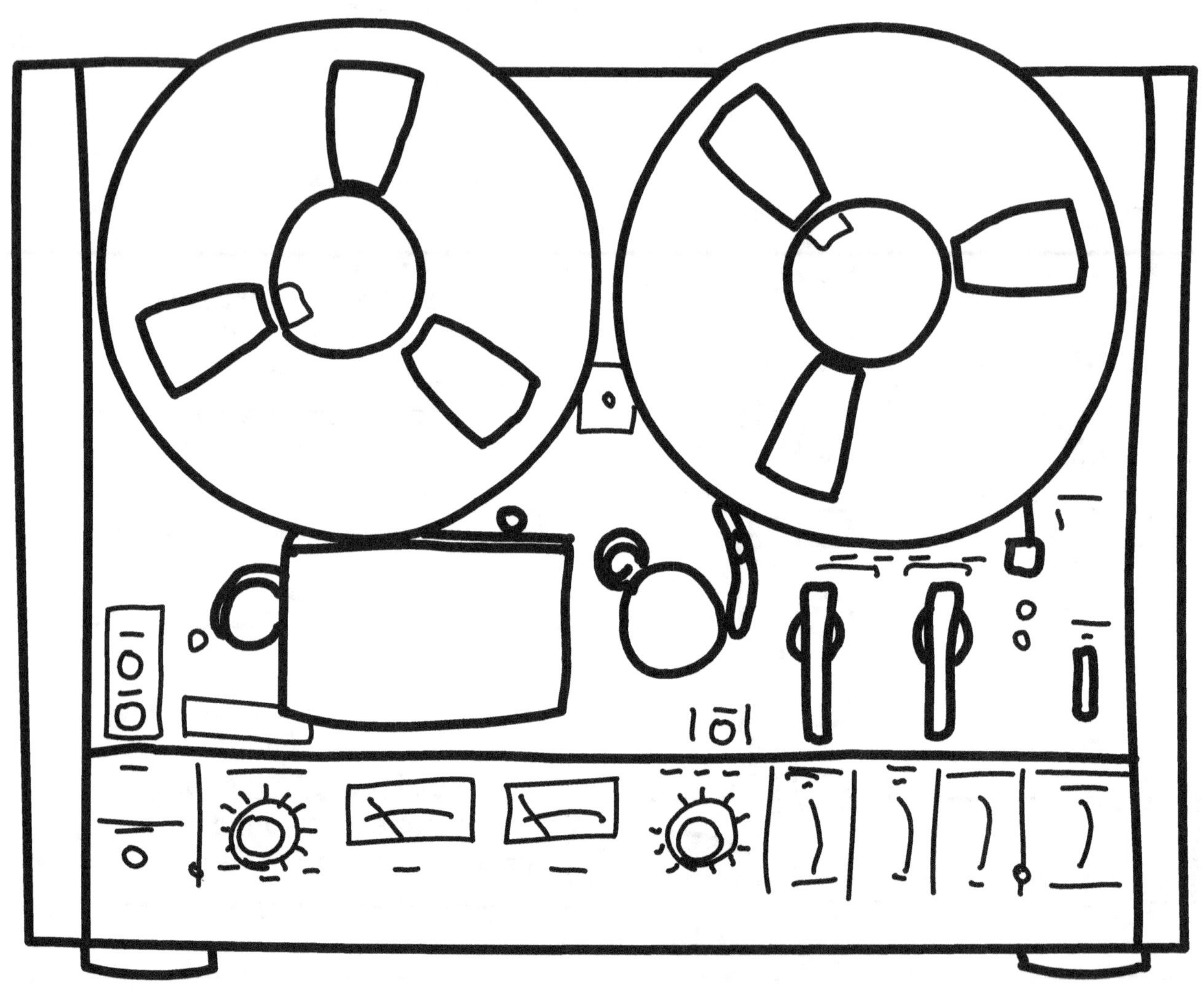

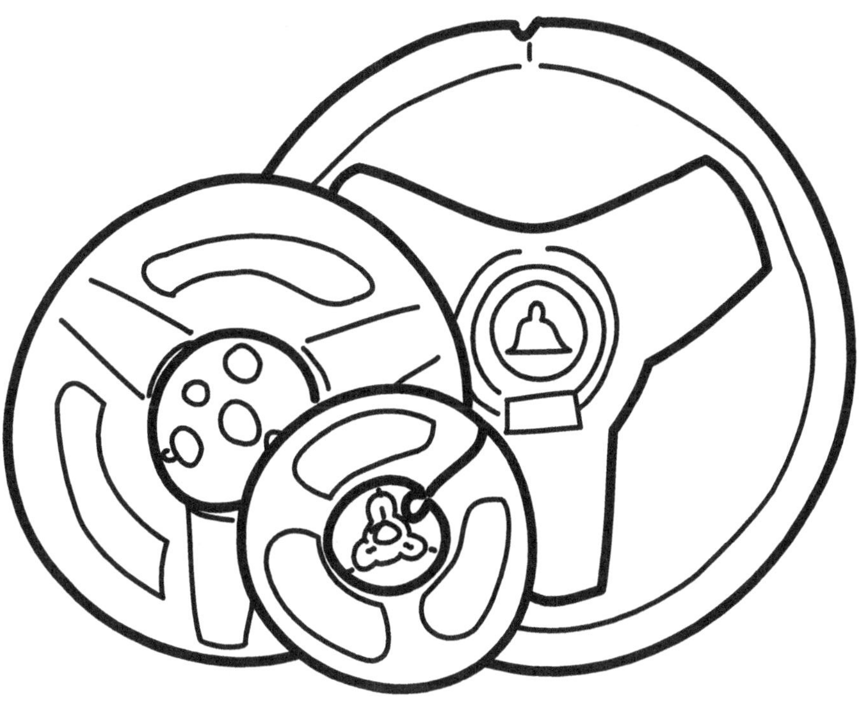

Fun fact
This format is
available in
many widths:
1/4", 1/2" 1"
& 2" are the
most common

Fun fact
This was the
primary format used
by professional
recording studios
until the late 1980s

Fun fact
A common tape speed is
7 1/2 inches per second (ips);
other recording speeds
are 3 3/4, 15, & 30 ips

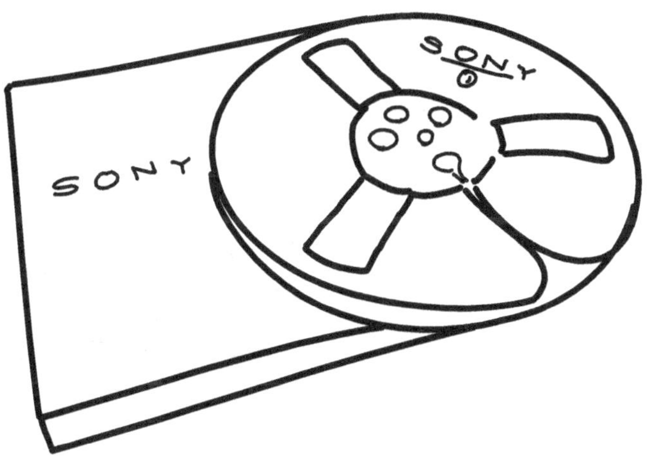

SONY
SONY

Tefifon

Developed by
Sony

Capacity
Small: 18 minutes
Medium: 1 hour
Large: 4 hours

Fun fact
This format used an endless loop plastic band onto which grooves were engraved and played back with a stylus

AKA
Tefi

Size
Small: 4.5 × 8 × 8.5 cm
Medium: 4.5 × 9.6 × 11.2 cm
Large: 4.5 × 13.5 × 15.7 cm

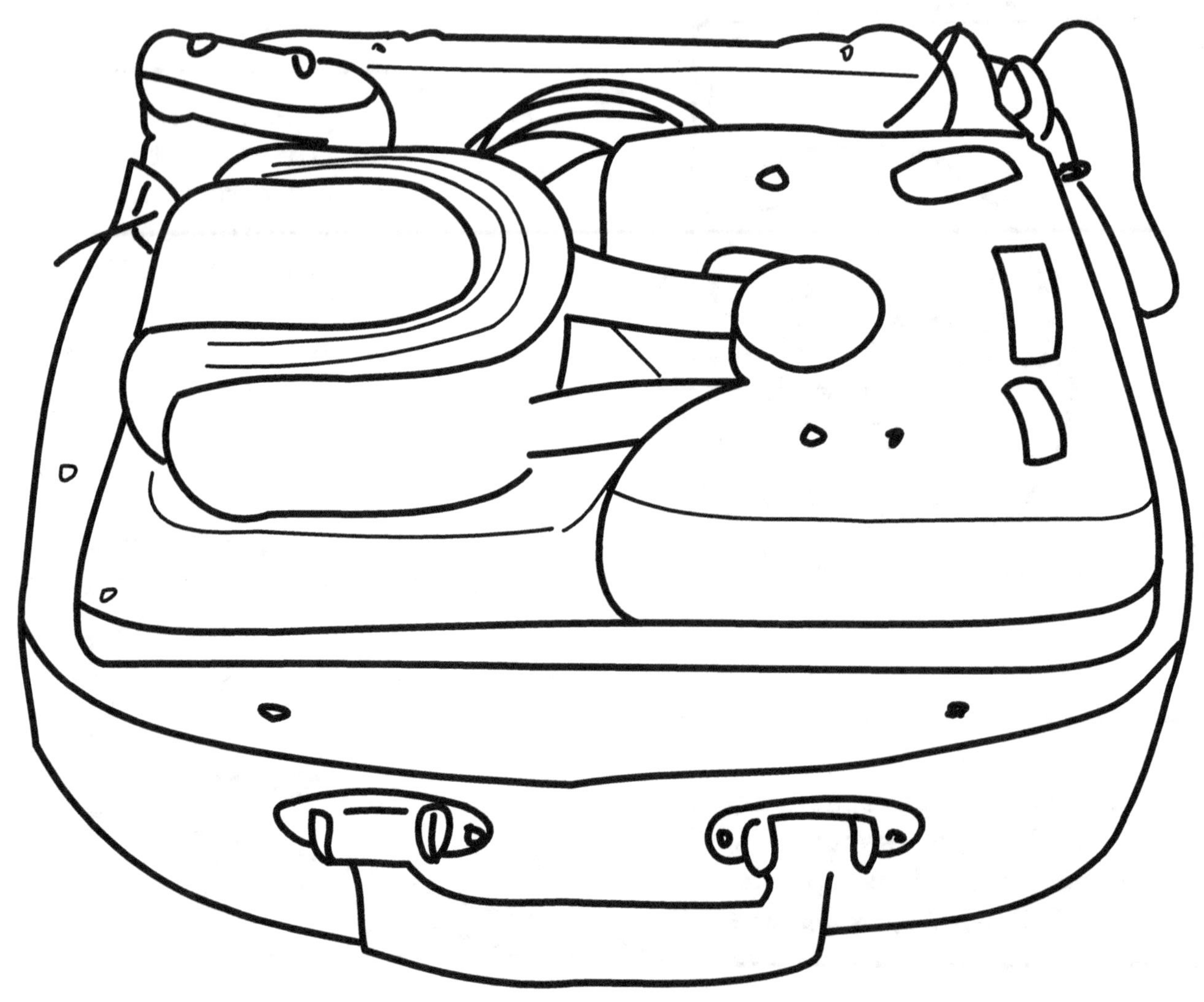

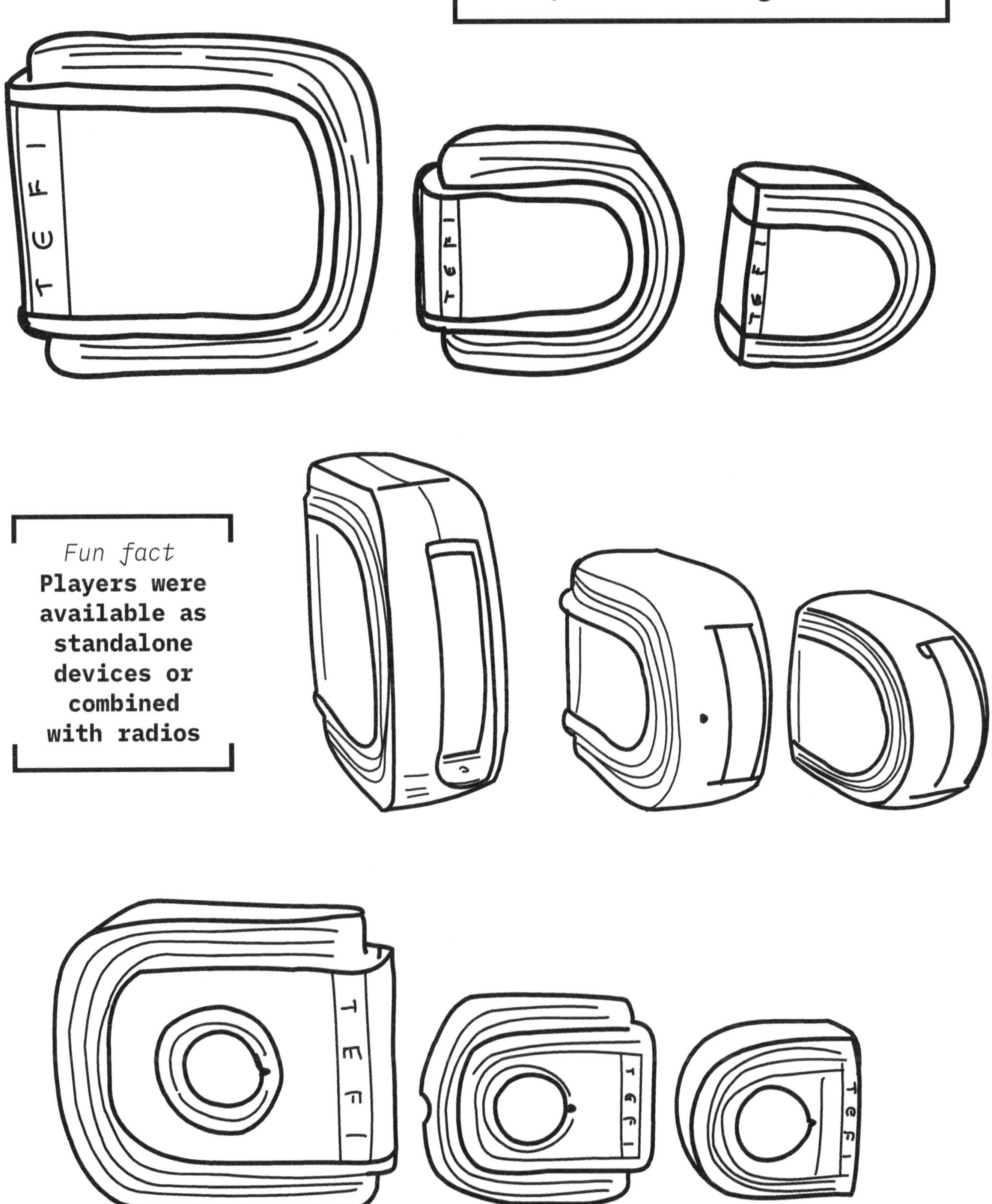

Fun fact
This format's sound quality
was higher than Gramophone
discs, but not Microgroove discs

Fun fact
Players were
available as
standalone
devices or
combined
with radios

TEFI

Dictabelt

Era
1947–1980

Developed by
American Dictaphone

Size
8.9 × 30 × 0.013 cm

Capacity
Standard speed: 15 minutes
Half speed: 30 minutes

AKA
**Dictaphone,
Memobelt**

Fun fact
Dictabelts were
red until 1964,
blue from
1964–1975,
then purple
until they were
discontinued

Fun fact
This format
was more
convenient and
had better audio
quality than
reusable wax
cylinders

Fun fact
This format
could be
folded and
fit into
a standard
letter-sized
envelope

Microgroove disc

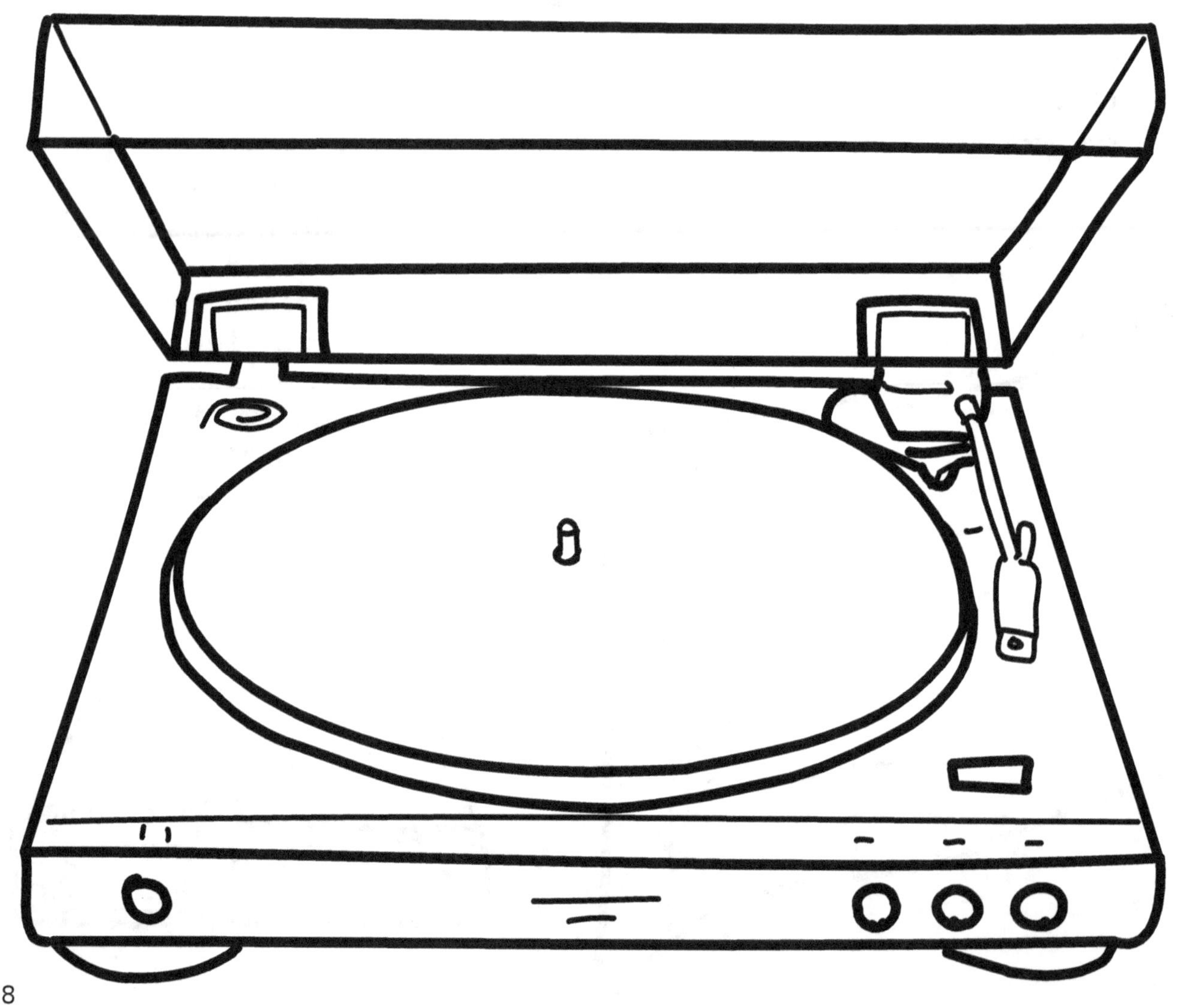

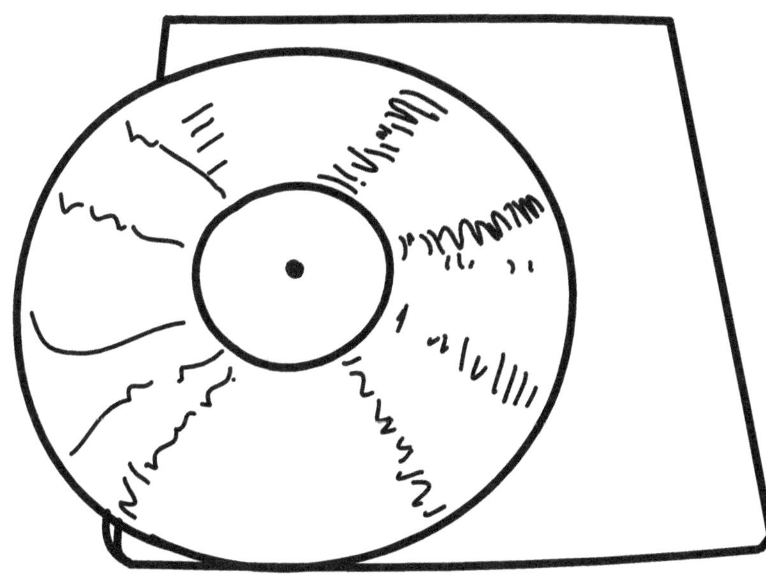

Fun fact
The most
well-known
version of
this format
today is
a 12" disc
played at
33 1/3 rpm

Fun fact
An earlier version of
this format was created by
RCA Victor in 1931, intended
to use as transcription discs

Fun fact
This format is
an updated verson
of the gramophone
disc and was
adopted as the
new standard by
the entire
record industry

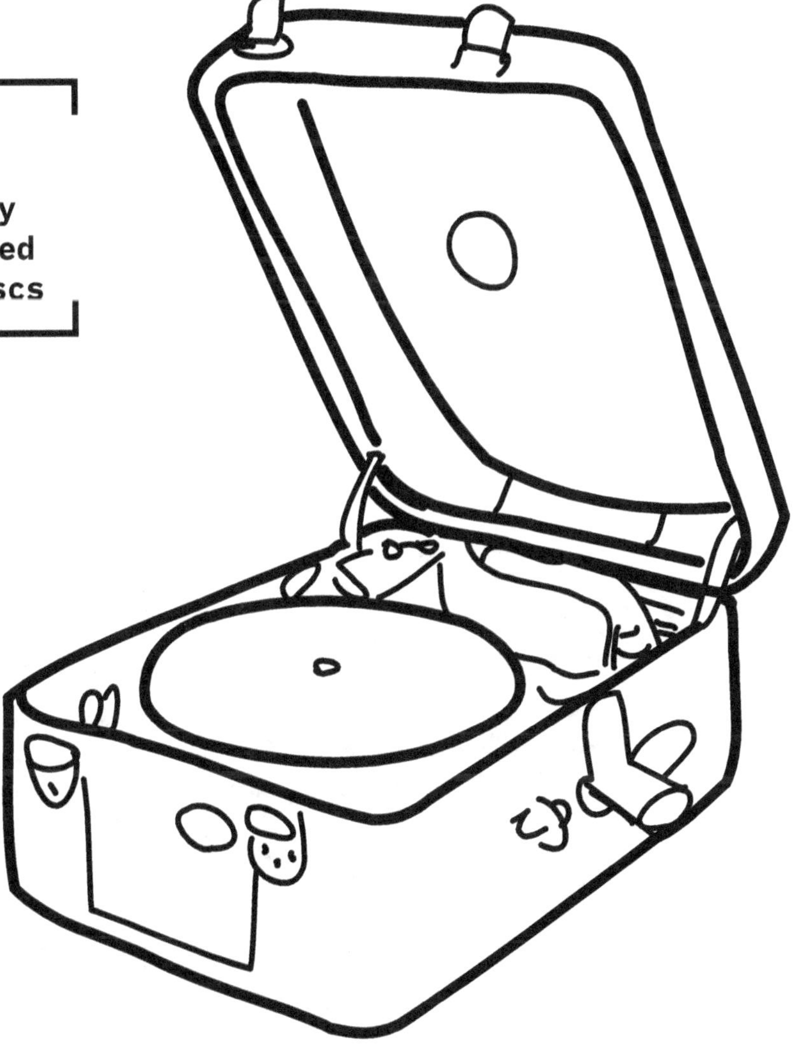

Minifon wire

Fun fact
This format
was used by
state government
agencies for
covert recording

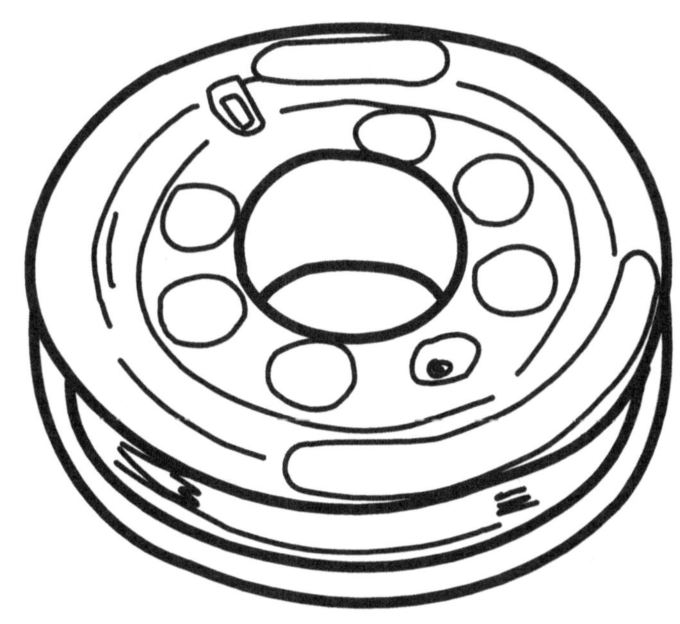

Fun fact
This format came
with accessories
like microphones
disguised as lapels
or watches

Fun fact
Including
battery
and spools,
this
format
weighed
less than
28 ounces

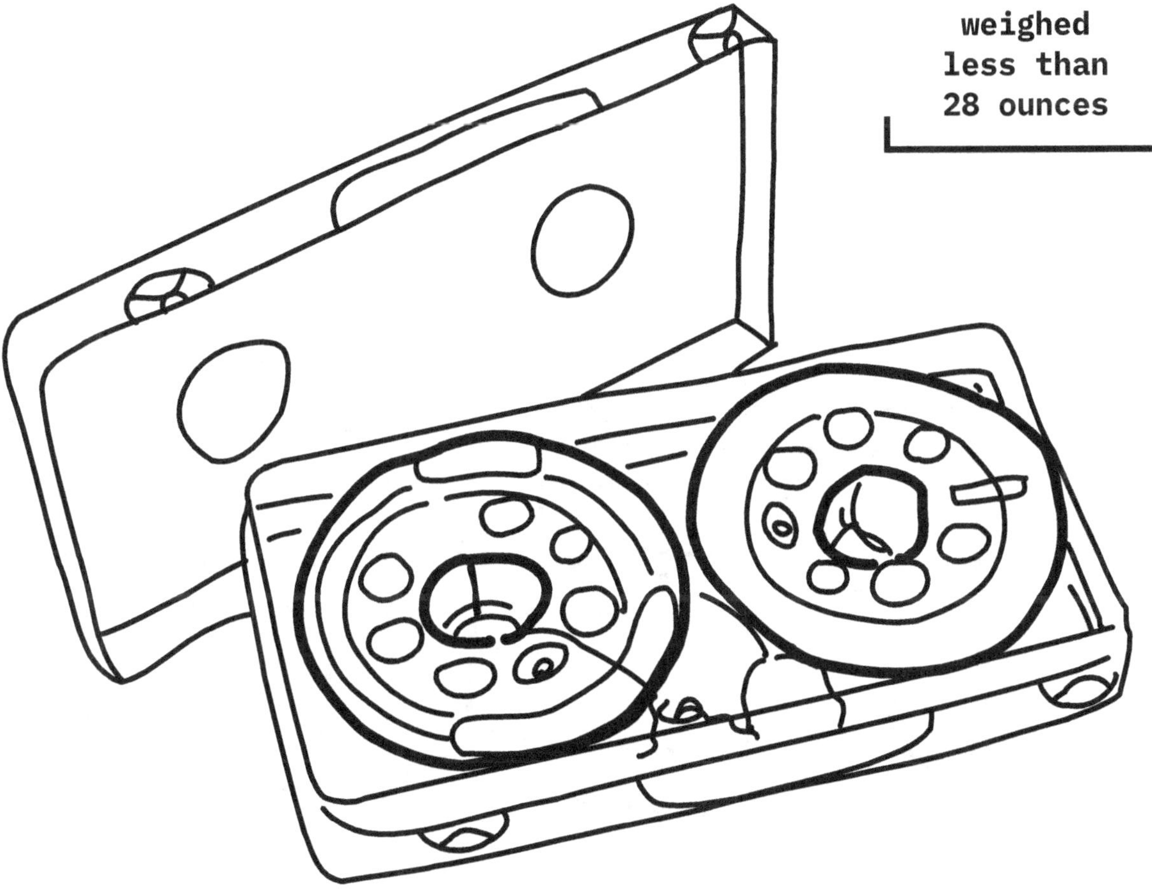

Fidelipac

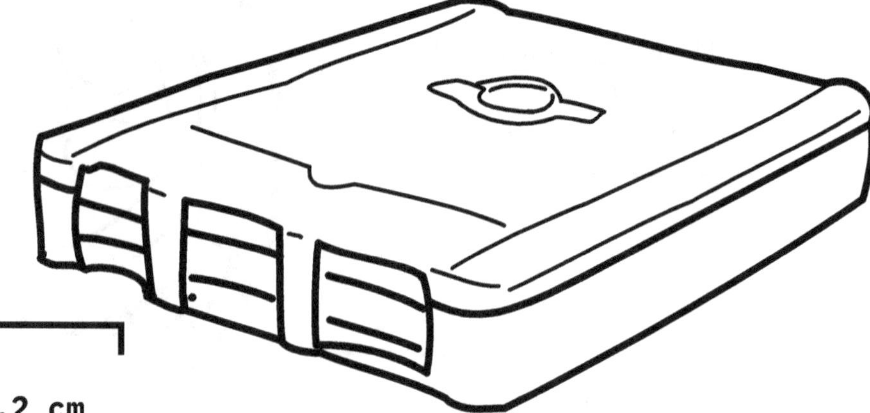

Capacity
Size A: 10 minutes
Size B: 20 minutes
Size C: 30 minutes

Size
Size A: 10.1 × 13.3 × 2.2 cm
Size B: 15.2 × 17.8 × 2.2 cm
Size C: 19.4 × 21.6 × 2.2 cm

AKA
**NAB cartridge,
cart tape**

Era
1954–1990s

Fun fact
**This format was used in
broadcasting for playback
of material such as radio
commercials or jingles**

Developed by
George Eash*

Fun fact
***The invention of this format is credited to George Eash as well as Vern Nolte (Automatic Tape Company) and is based on endless loop magnetic tape cartridge design by Bernard Cousino**

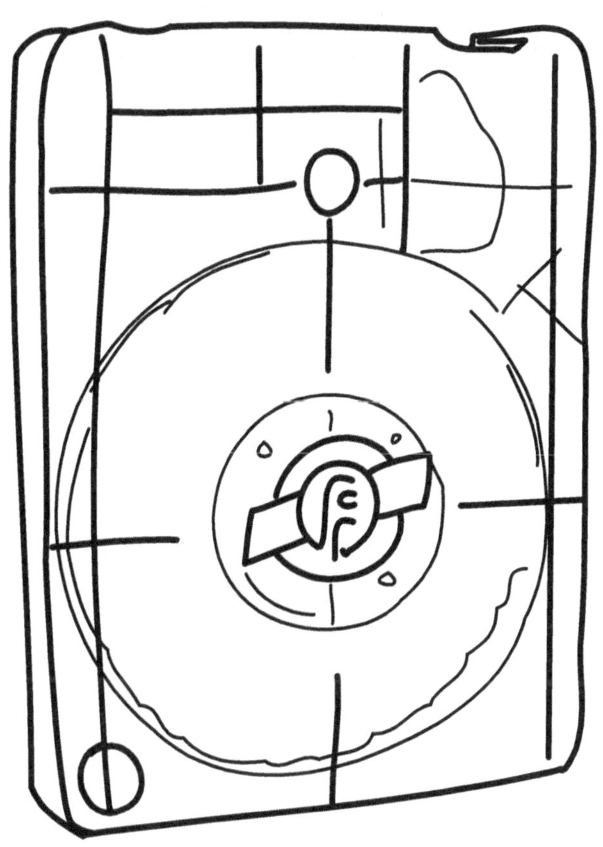

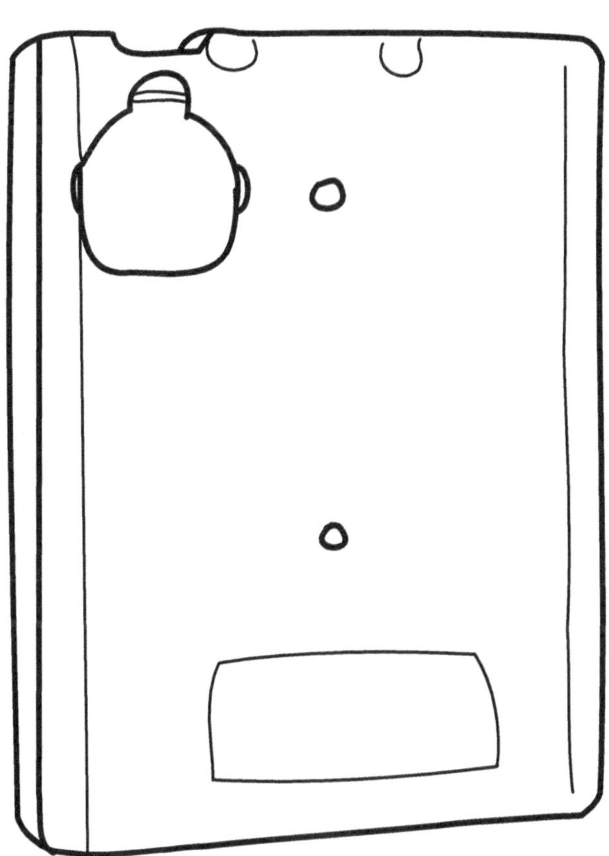

Fun fact
Size A was used for broadcasting commercials, and the B and C sizes were usually used for background music

Stenorette

Format
analog

Era
1954–1970s

Developed by
Grundig

Capacity
45 minutes

Size
3"

GRUNDIG
CASSETTE

stenorette

Fun fact
The initial release of this format was nicknamed "tree frog" because of its green color

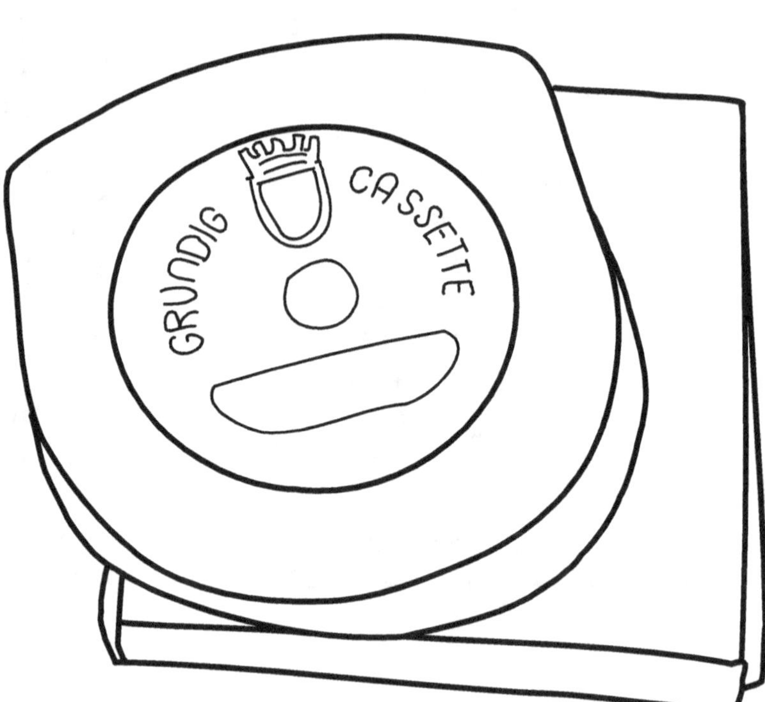

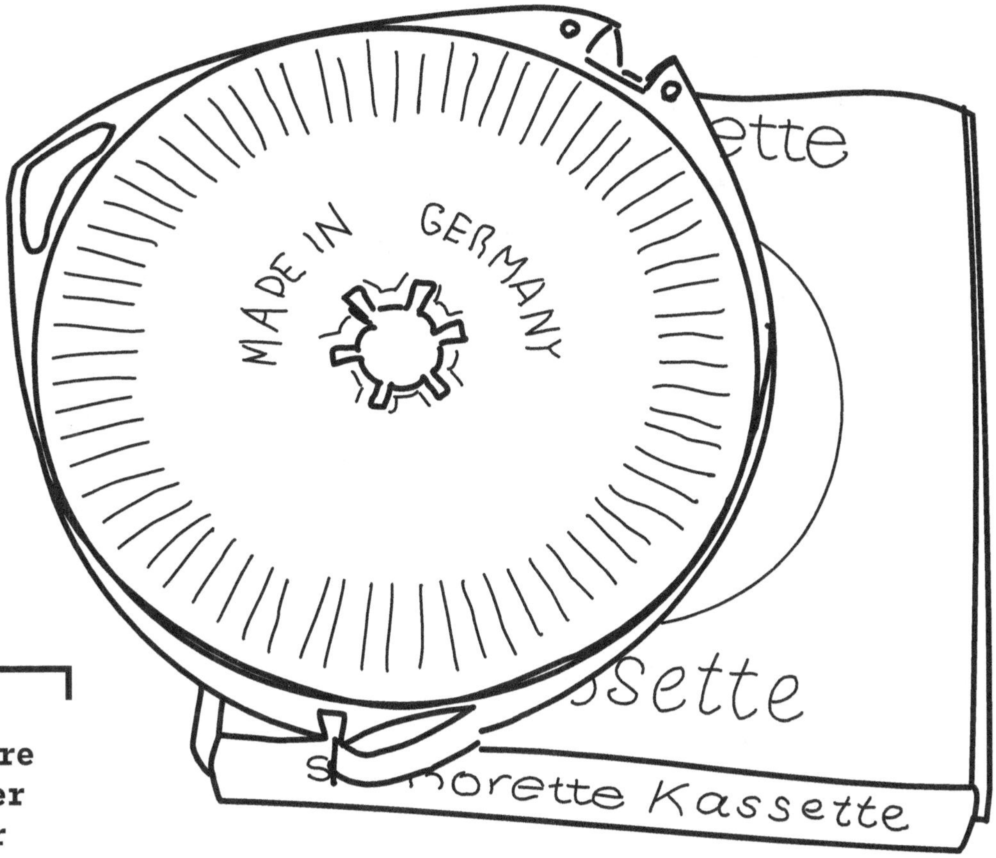

MADE IN GERMANY

Kassette

Stenorette Kassette

Cassettes were held together by a rubber retaining ring underneath the top cover

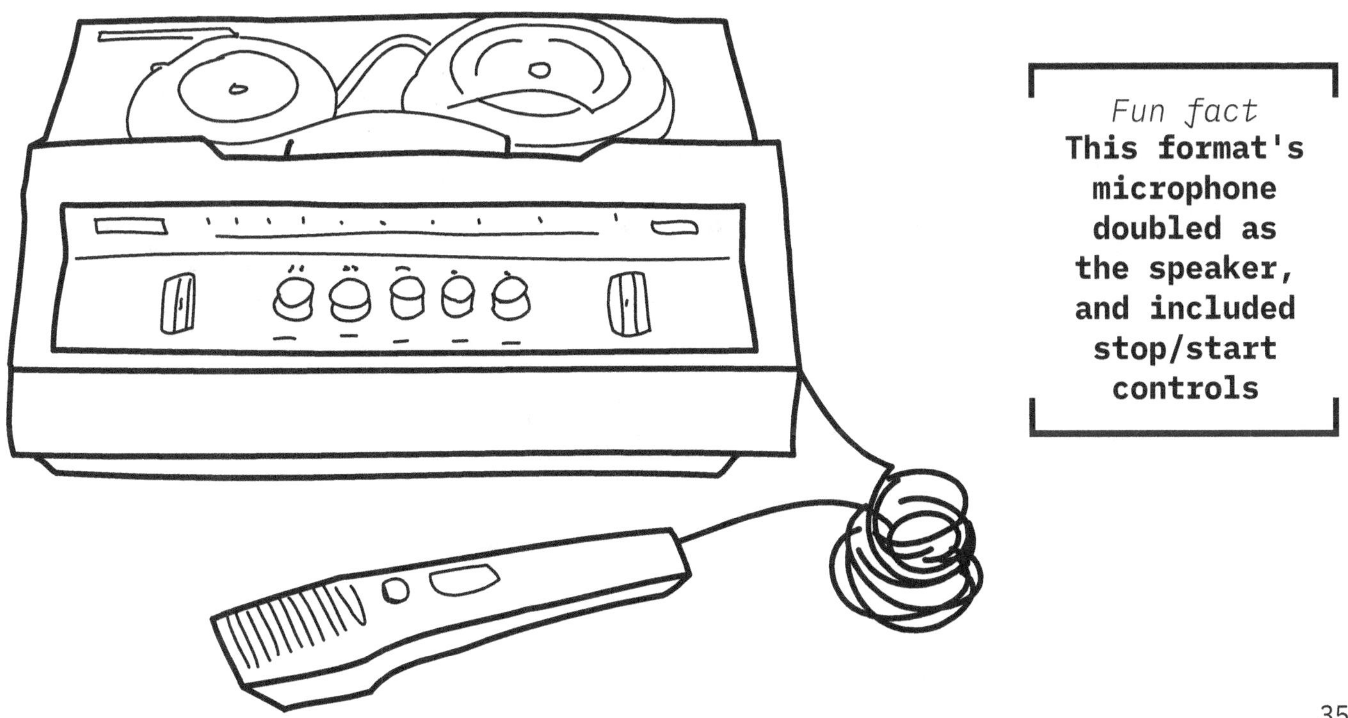

This format's microphone doubled as the speaker, and included stop/start controls

Sound Tape Cartridge

Era
1958–1964

Developed by
RCA

Size
13.7 × 19.7 × 1.3 cm

AKA
RCA Tape Cartridge

Capacity
30 minutes (per side)

Fun fact
This format was designed to be more convenient than open reel by avoiding the need to thread tape into a machine

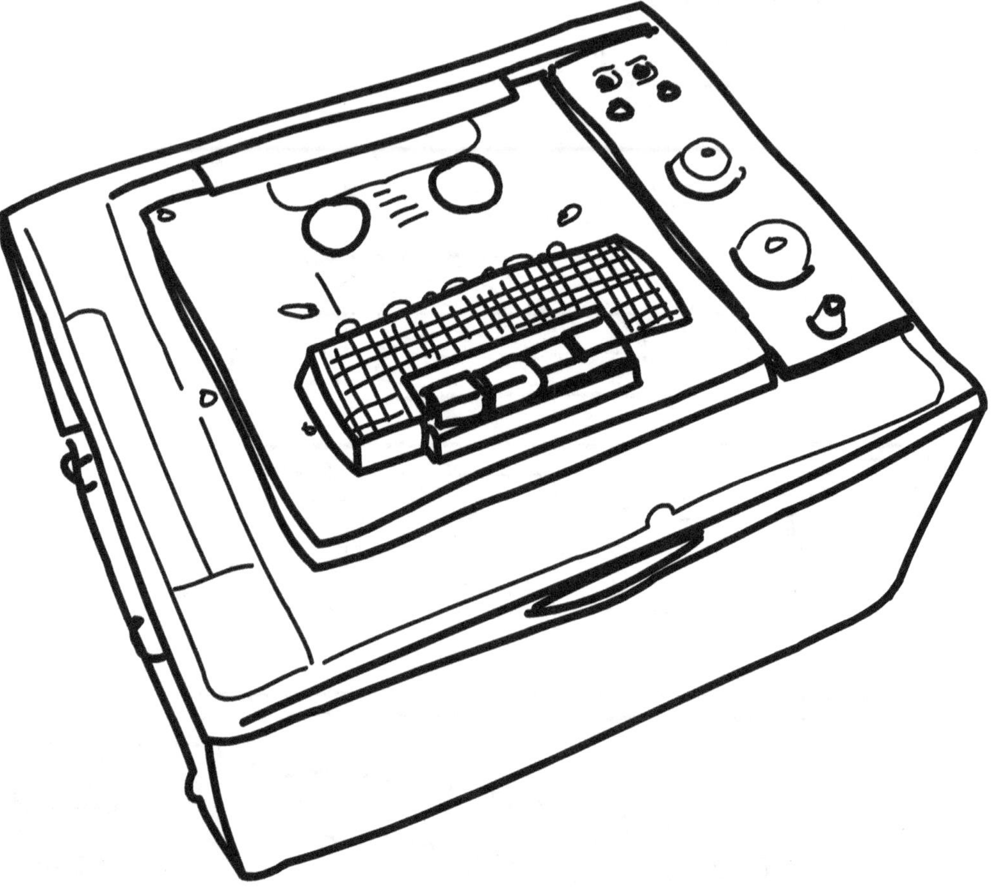

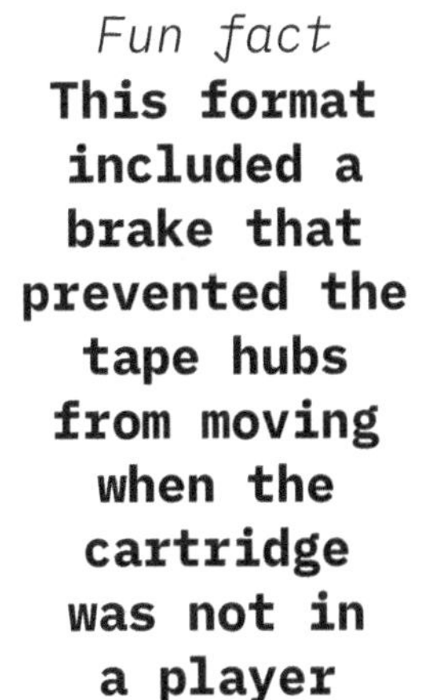

Fun fact
This format was introduced in 1958, following four years of development

Minifon tape

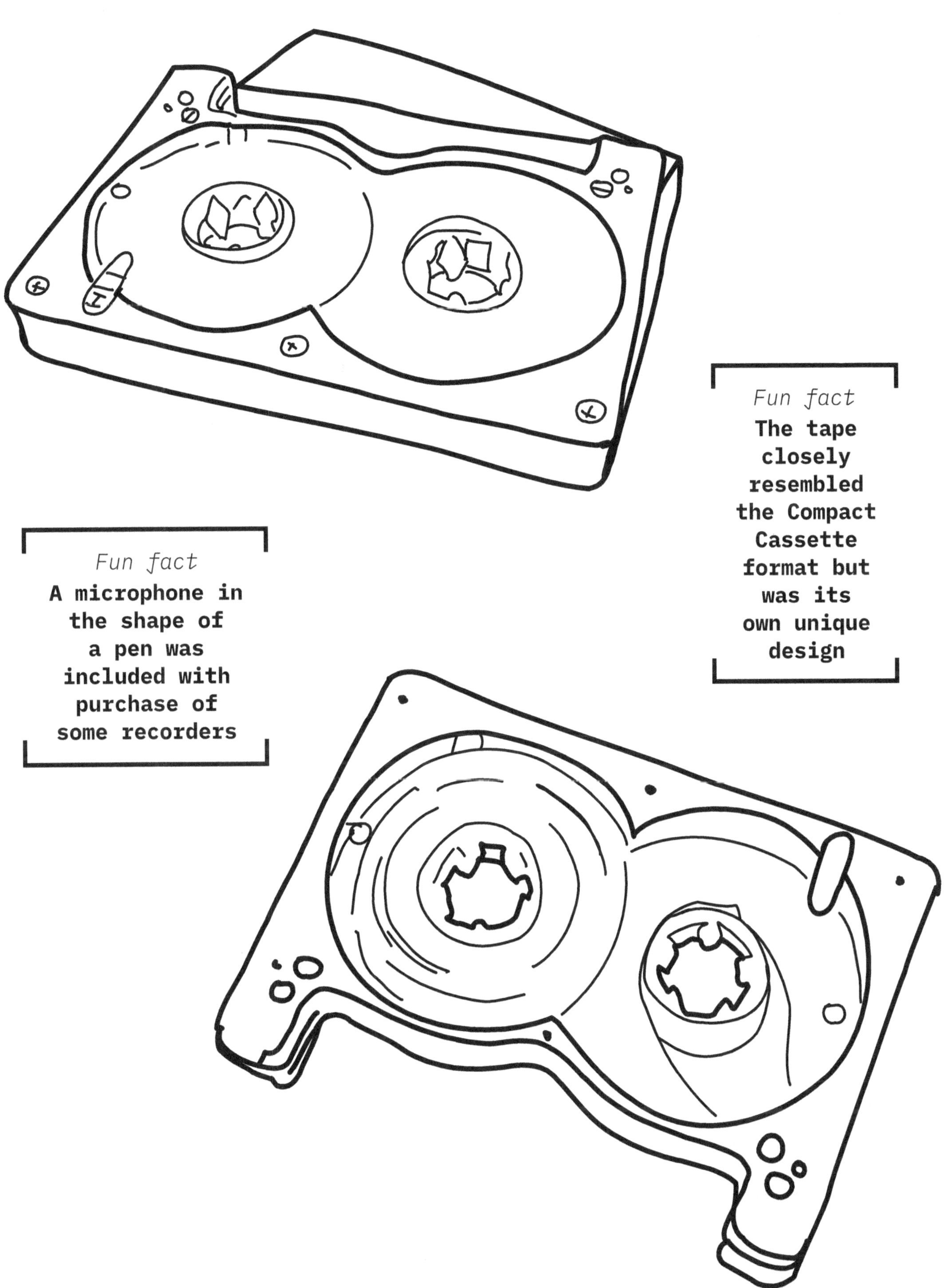

Fun fact
A microphone in
the shape of
a pen was
included with
purchase of
some recorders

Fun fact
The tape
closely
resembled
the Compact
Cassette
format but
was its
own unique
design

Stereo-Pak

AKA
**4-track,
CARtridges**

Era
1962–1970

Capacity
45 minutes

Size
13.3 × 10.2 × 2 cm

Developed by
Earl 'Madman' Muntz

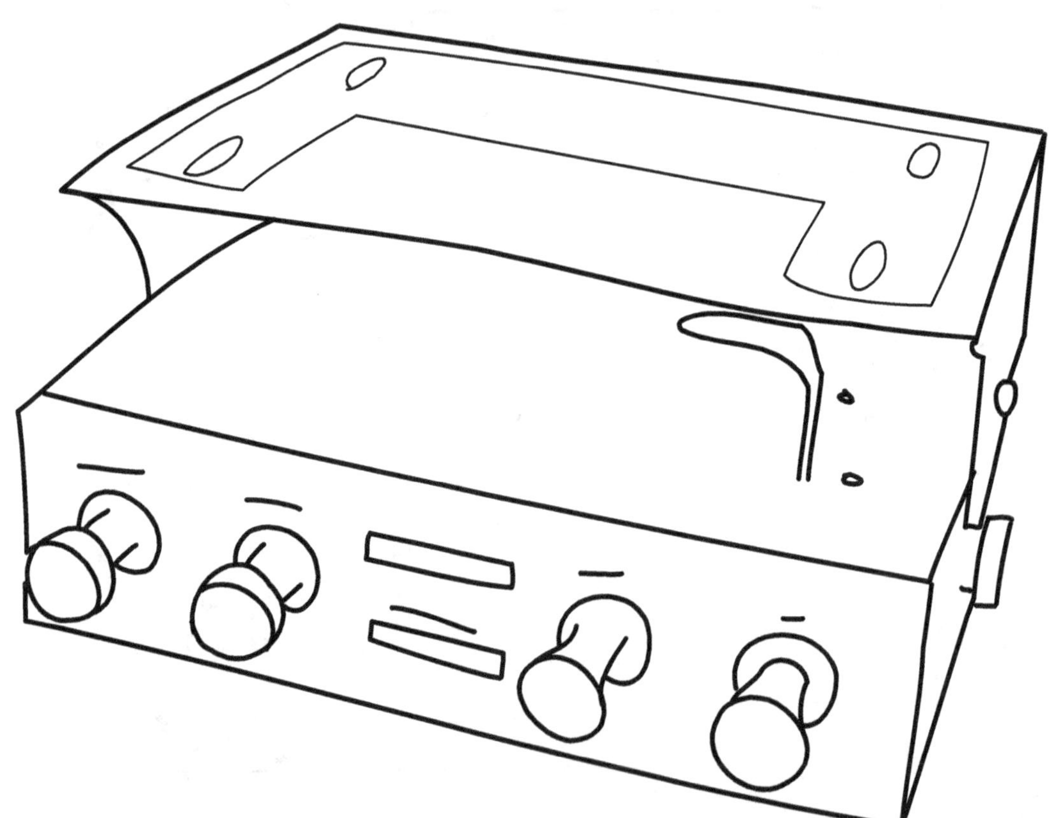

Fun fact
Unlike its
predecessor,
the Fidelipac,
this format used
a movable head
to switch
between two
programs

Fun fact
This format
was installed
in cars but
also available
as home players

Fun fact
This format
was a market
success for
several years
until its
successor,
the 8-Track,
became more
popular
despite being
lower quality

8-track

Developed by
Lear Industries

AKA
Stereo 8

Era
1963–1980s

Capacity
45 minutes

Format
analog

Size
13.3 × 10.2 × 2 cm

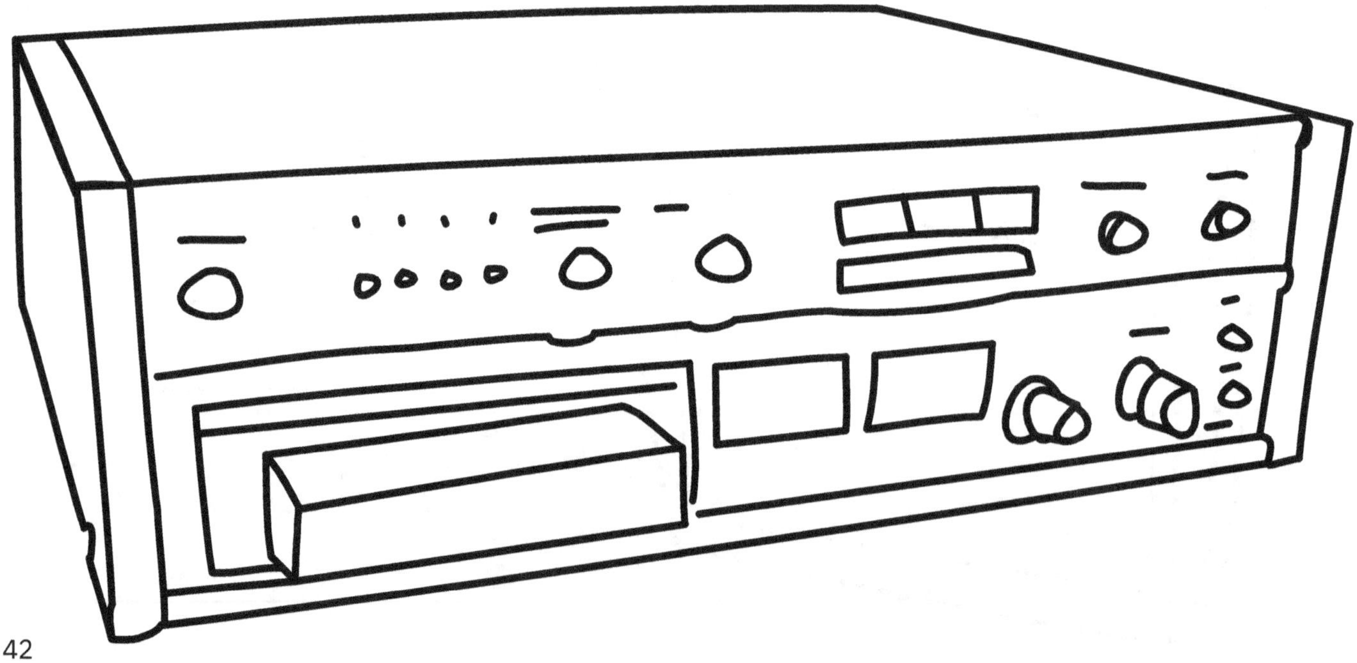

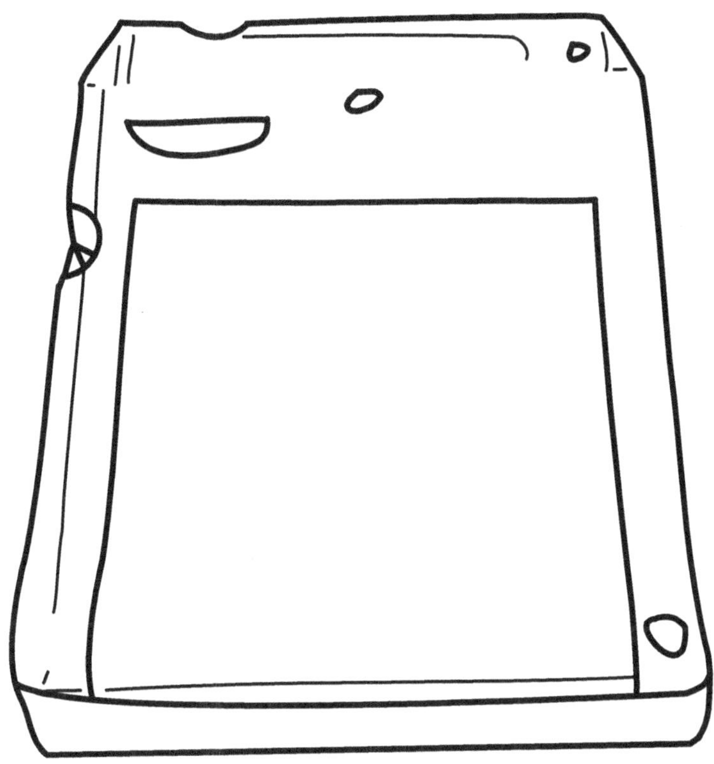

Fun fact
This format had
8 tracks for
4 stereo
programs, and
the format
could change
between the
4 options
by a device
that would
shift the
tape head

Fun fact
This format
was an
evolution
of the
Stereo-Pak
format

Fun fact
This format
worked as
an endless
loop and
it could
not be
rewound

Compact Cassette

Fun fact
This format's design won out over other competing cassette designs partly due to a decision to freely license the design after 1965, and the assurance that other companies would support the format

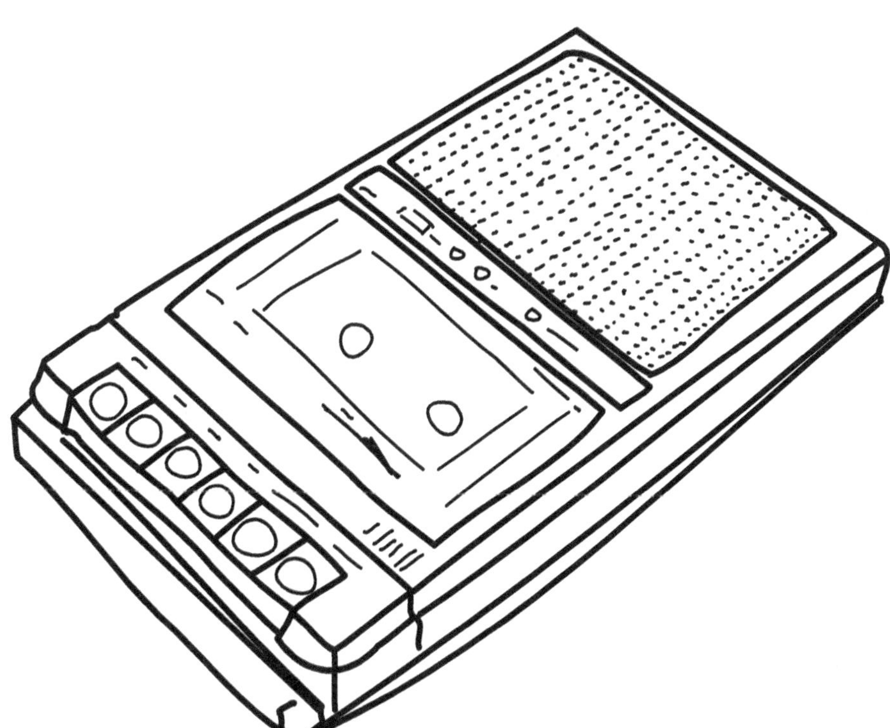

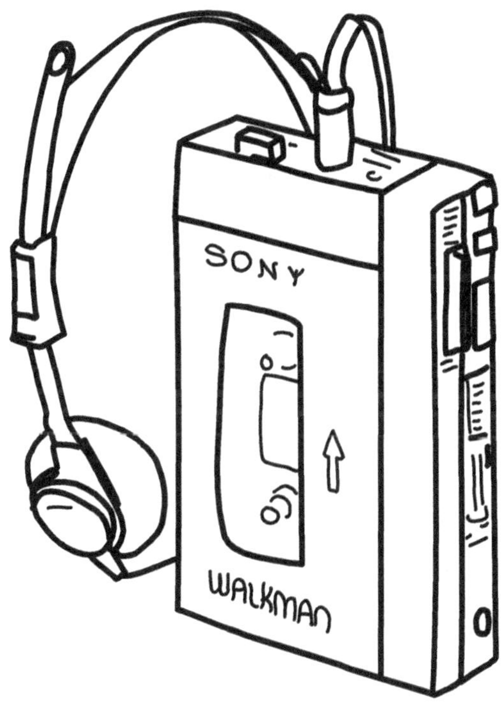

SONY
WALKMAN

Fun fact
In addition to audio, this format was used as data storage for microcomputers in the late 1970s-1980s

Fun fact
This format was initially called the Pocket Recorder; the name Compact Cassette wasn't used until around 1966

Micro Pack 35

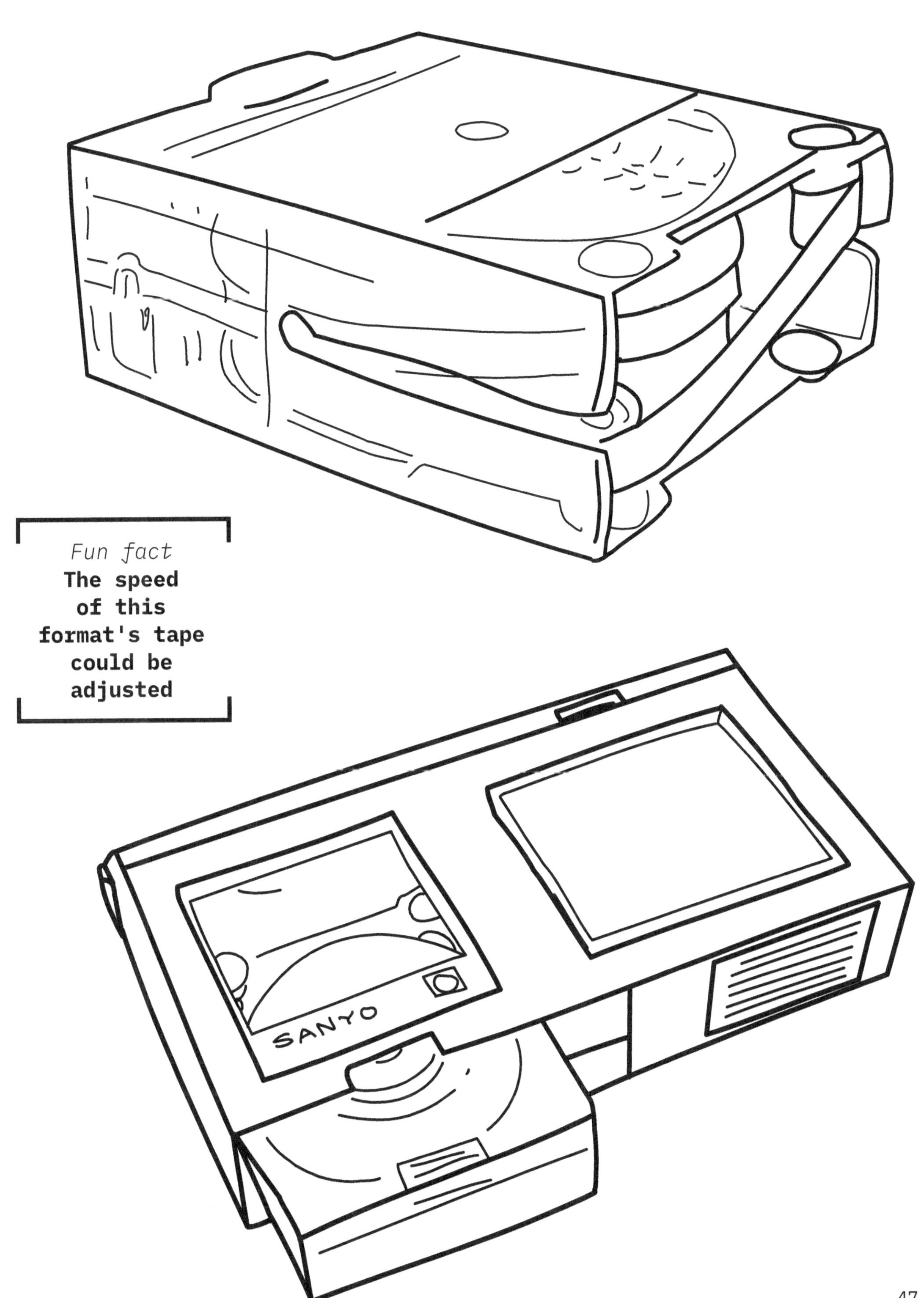

SANYO

Fun fact
The speed
of this
format's tape
could be
adjusted

Sabamobil

Size
6.6 cm × 7.4 cm × 4.8 cm

Capacity
20 minutes (per side)

Developed by
Sanyo

Fun fact
This format was introduced a year after the Compact Cassette and 8-track, and wasn't able to compete in the market

Era
1964–1970

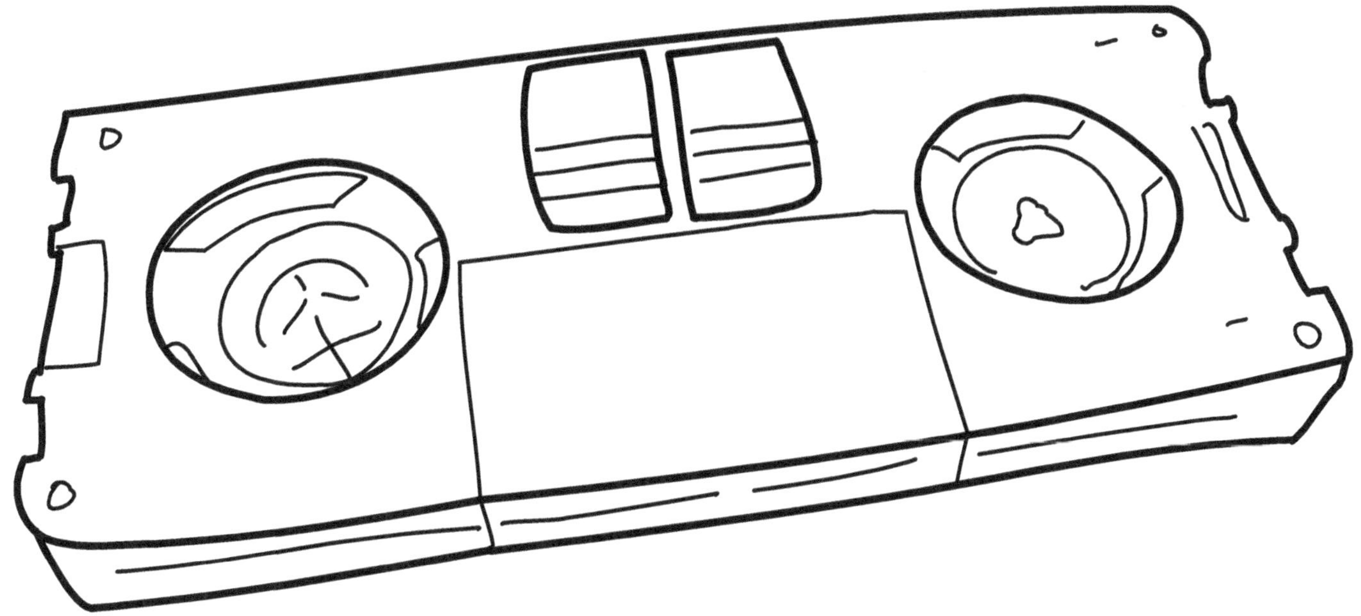

Fun fact
**The cartridge
could be
opened by
removing the
two holding
clamps**

Fun fact
**One model,
the TK-R12,
had a built-in
radio and could
be portable,
using 5 D-size
batteries**

DC-International

Fun fact
DC is short for double cassette
because it was a two-reel cassette

Fun fact
This format
could enable
write protection
by putting an
insert into a
recess in the
base of the
cassette

PlayTape

Developed by
Frank Stanton

Era
1966–1970

Capacity
24 minutes

Size
7 × 8.5 × 1.2 cm

Fun fact
**PlayTape were issued
in different colors
according to the content:
Red cartridge (two songs)
Black cartridge (four songs)
White cartridge (eight songs)
Blue cartridge (children)
Gray cartridge (educational)**

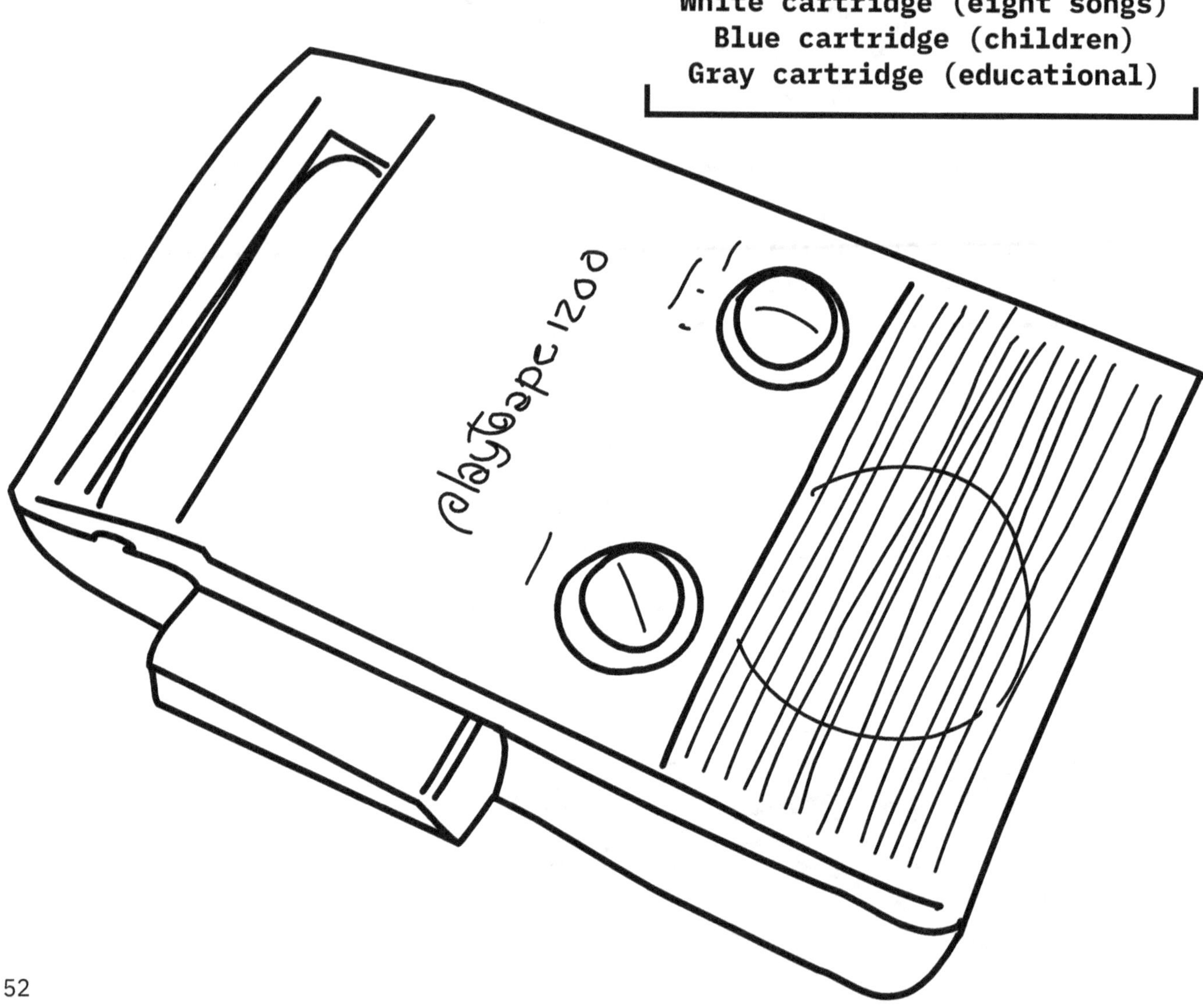

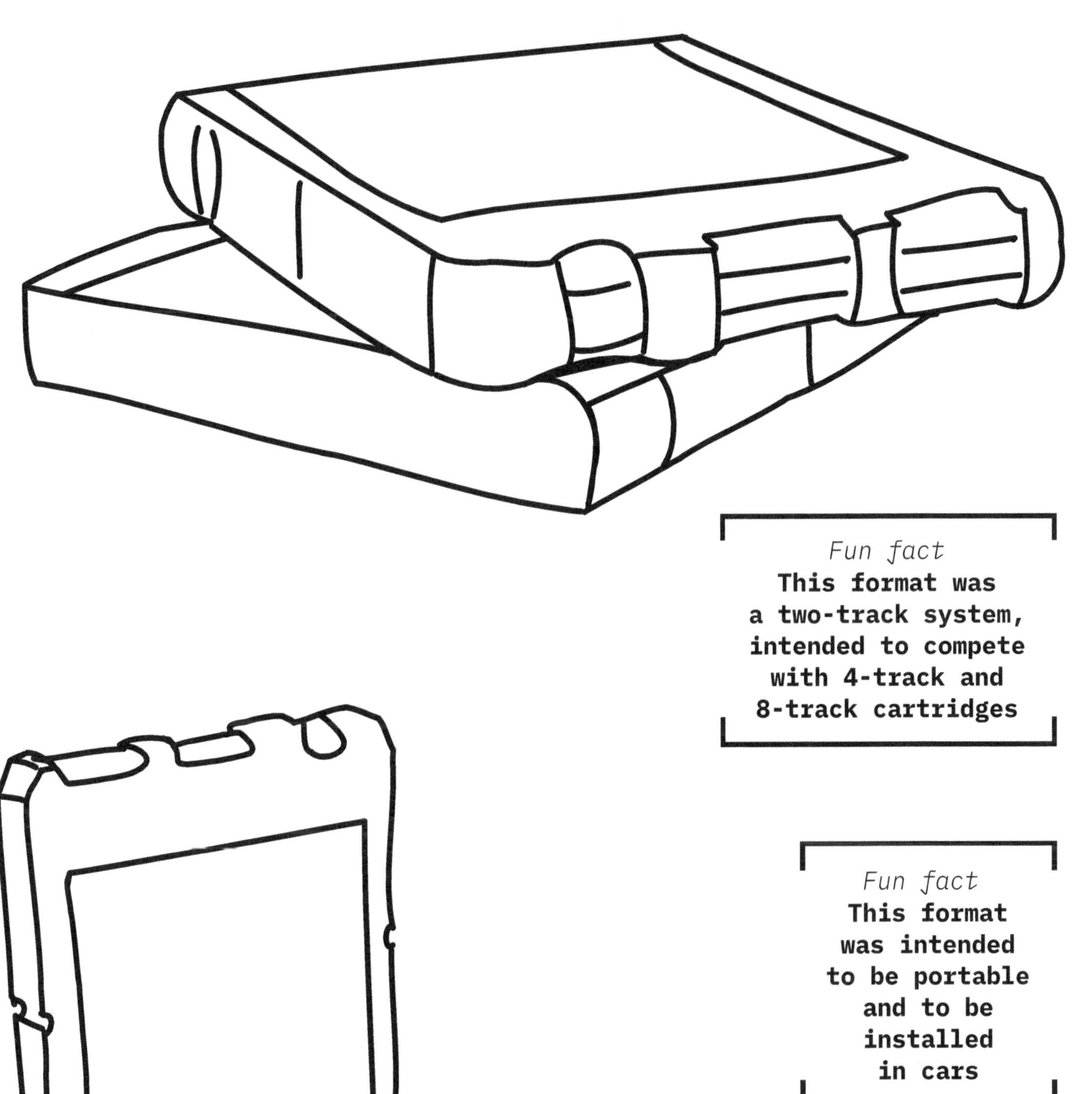

Fun fact

This format was a two-track system, intended to compete with 4-track and 8-track cartridges

Fun fact

This format was intended to be portable and to be installed in cars

Mini-cassette

Fun fact
There was a smaller
version of this
format called an
Ultra Mini-Cassette
that could record up
to 10 minutes on
each side

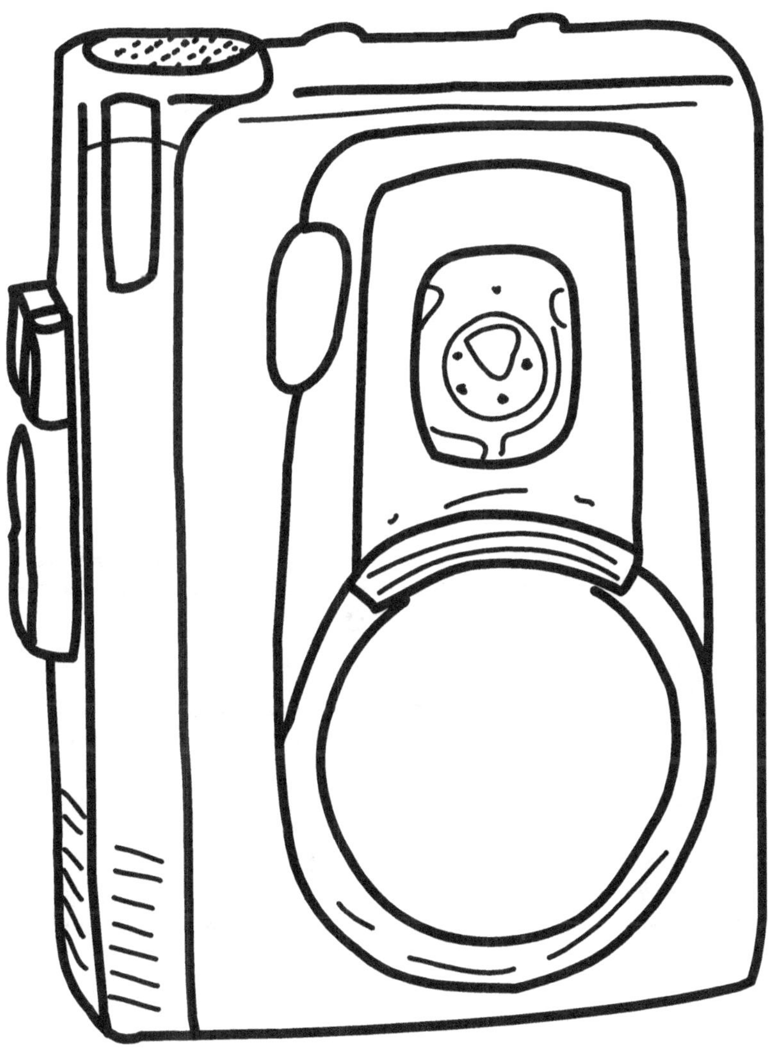

Fun fact
This format
was mostly
used in
dictation
machines
but could
also be used
as data
storage

Microcassette

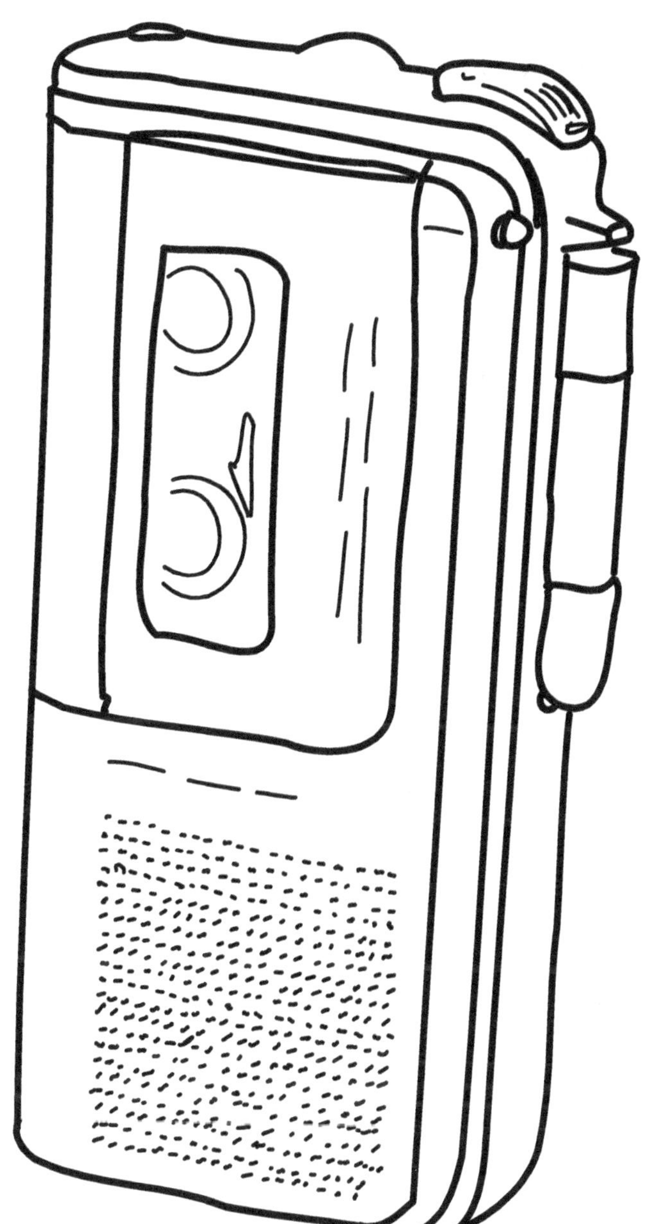

Fun fact
This format
was slightly
more narrow
(by 5mm)
than the
Mini-cassette

Fun fact
This format
was commonly
used in
dictation
machines or
telephone
answering
machines,
but also
computer
data storage

Fun fact
This format
has the same
width of
magnetic tape
as the
Compact
Cassette
but in a
container
roughly
one quarter
the size

HIPAC

Format	Size
analog	**7 × 8.5 × 1.2 cm**

Developed by	Era	AKA	Capacity
Pioneer	**1971–1973**	**HiPac**	**60 minutes**

Fun fact
This format is a successor of the PlayTape cartridge, licensed by Toshiba

Fun fact
This format was intended for use in vehicles but adoption was unsuccessful

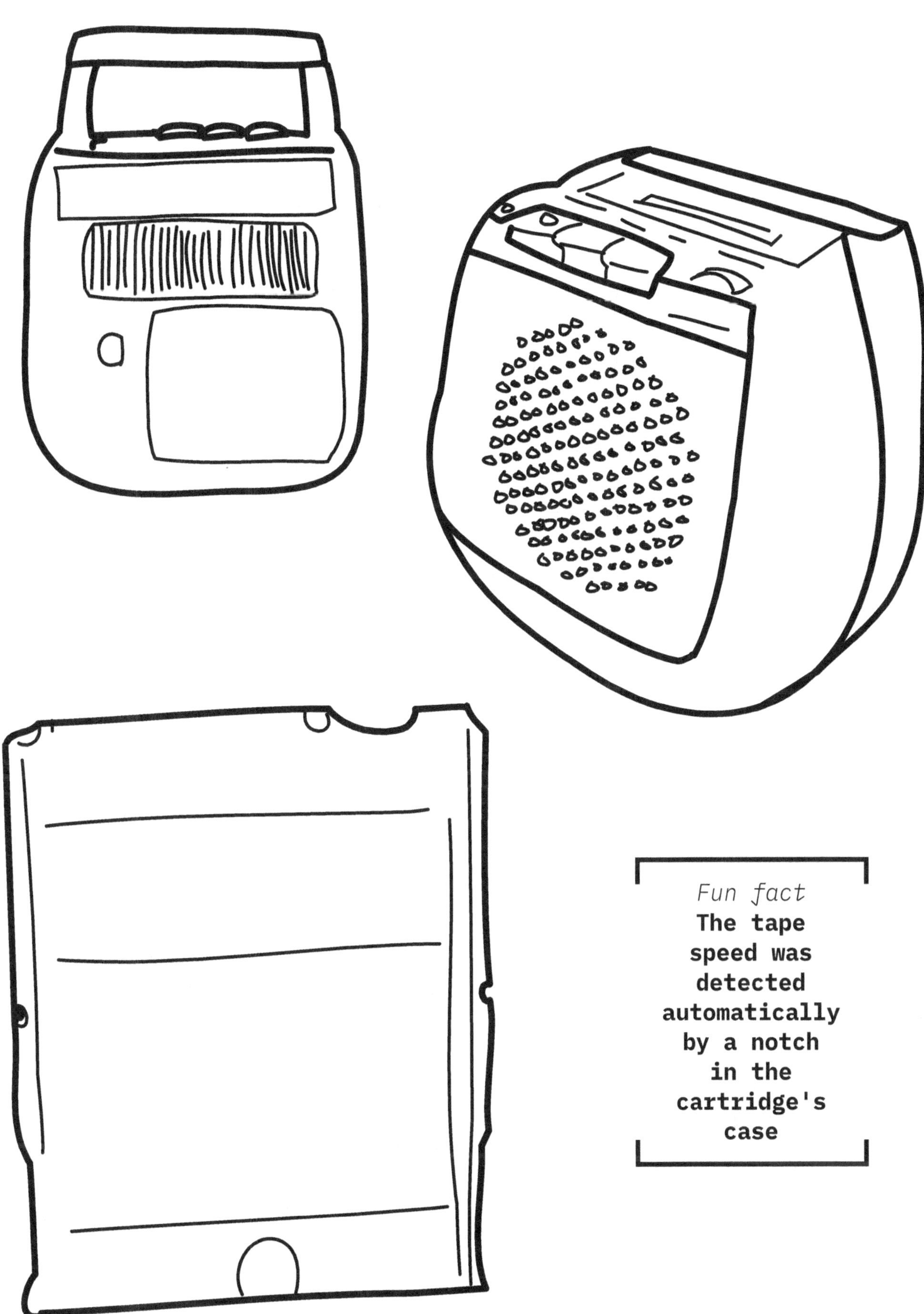

Fun fact
The tape
speed was
detected
automatically
by a notch
in the
cartridge's
case

3/4" U-matic

Format
digital

Era
1971–1990s

Developed by
Sony

Capacity
Small: 20 minutes
Standard: 60 minutes

Size
Small (S): 18.4 × 12.2 × 3.2 cm
Standard (SP): 22.1 × 14.0 × 3.2 cm

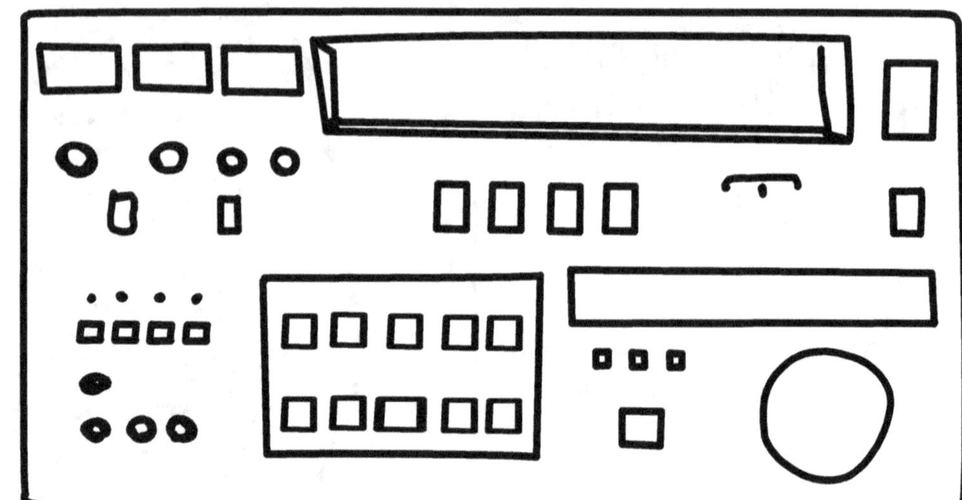

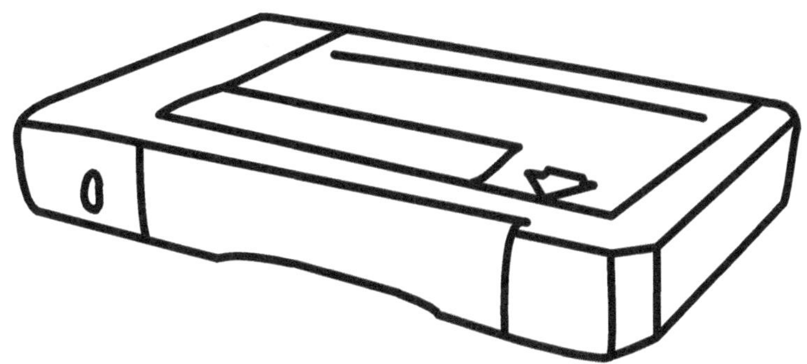

Fun fact
While this format was primarily used to store analog video, it was also used for the storage of digital audio data

Fun fact
The 44.1 kHz sampling standard used in Compact Discs was defined by the bandwidth available this format's tapes

Fun fact
All tapes have a round red button on the back that can be removed to prevent accidental recording

Elcaset

AKA
L-cassette

Capacity
90 minutes

Fun fact
This format's name means L-cassette, or large cassette, since the cartridge was roughly double the size of the Compact Cassette

Era
1976–1980

Size
15 × 10 × 2 cm

Fun fact
This format was meant to have the audio quality of reel-to-reel with the convenience of the Compact Cassette

Developed by
Panasonic, Teac, Sony

Fun fact
This format was largely a market failure

LaserDisc

AKA
LD, MCA DiscoVision

Format
analog/digital

Era
1978–2001

Size
12" discs

Capacity
1 hour
(per side)

Developed by
Philips, MCA Inc.,
Pioneer Corporation

LaserDisc

Fun fact
In the UK,
"LaserDisc" was used
for discs that
hold digital audio
and "LaserVision" was
used for discs with
analog audio and video

Fun fact
Audio stored
on this format
could be analog
or digital; video
could only be analog

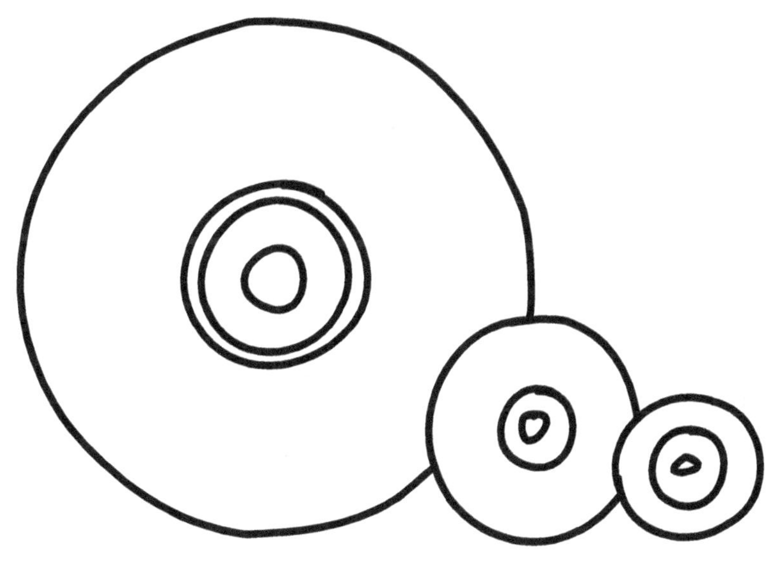

Fun fact
This format's digital
audio quality was
better than anything
else available to
consumers at the
time, but analog
audio quality was
inconsistent

Compact Disc

Developed by
Philips & Sony

AKA
Compact Disc
Digital Audio, CD

Capacity
80 minutes

Fun fact
Tracks on this format begin
on the inside of the disc
and spiral outward

Era
1982–

Size
4.7" discs

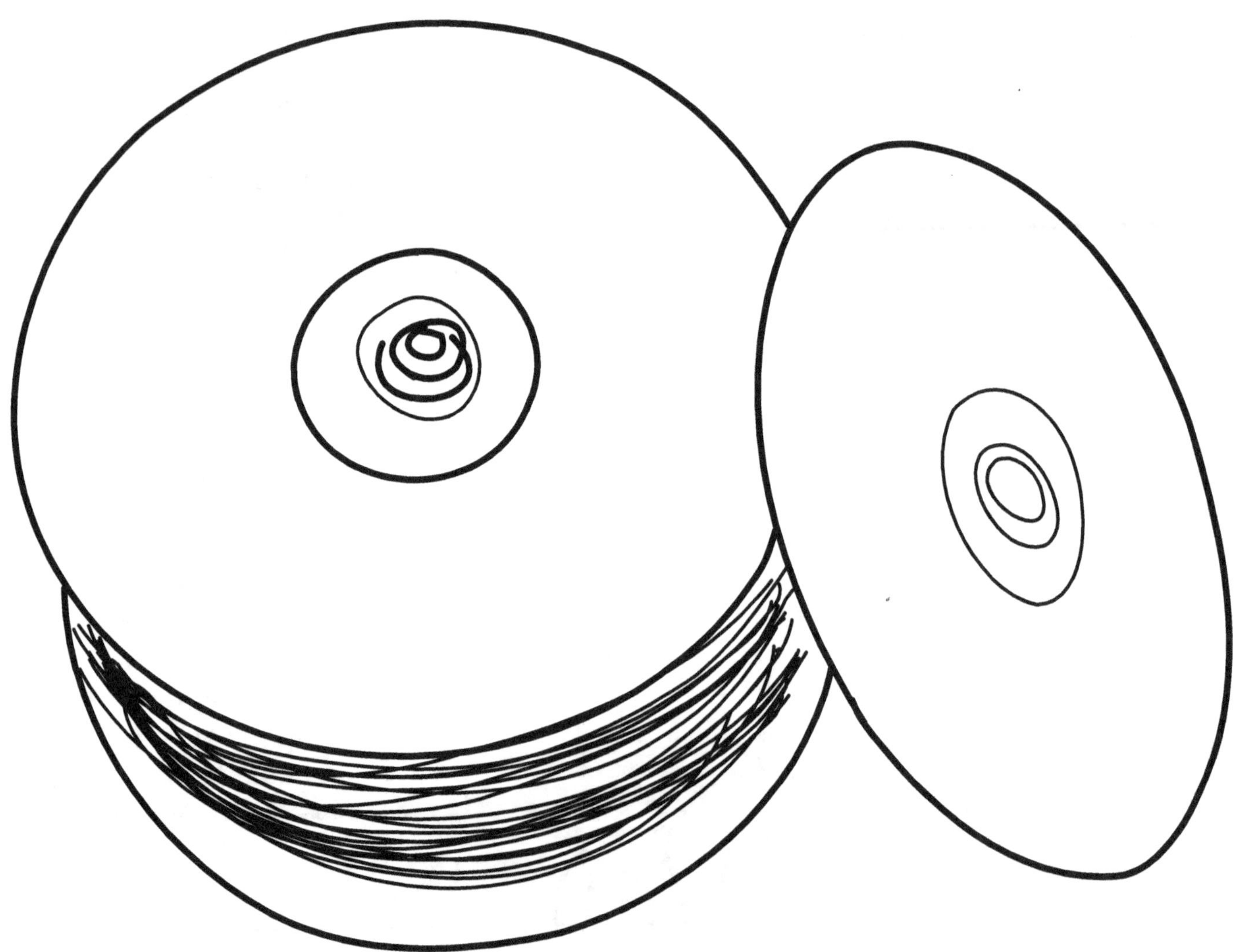

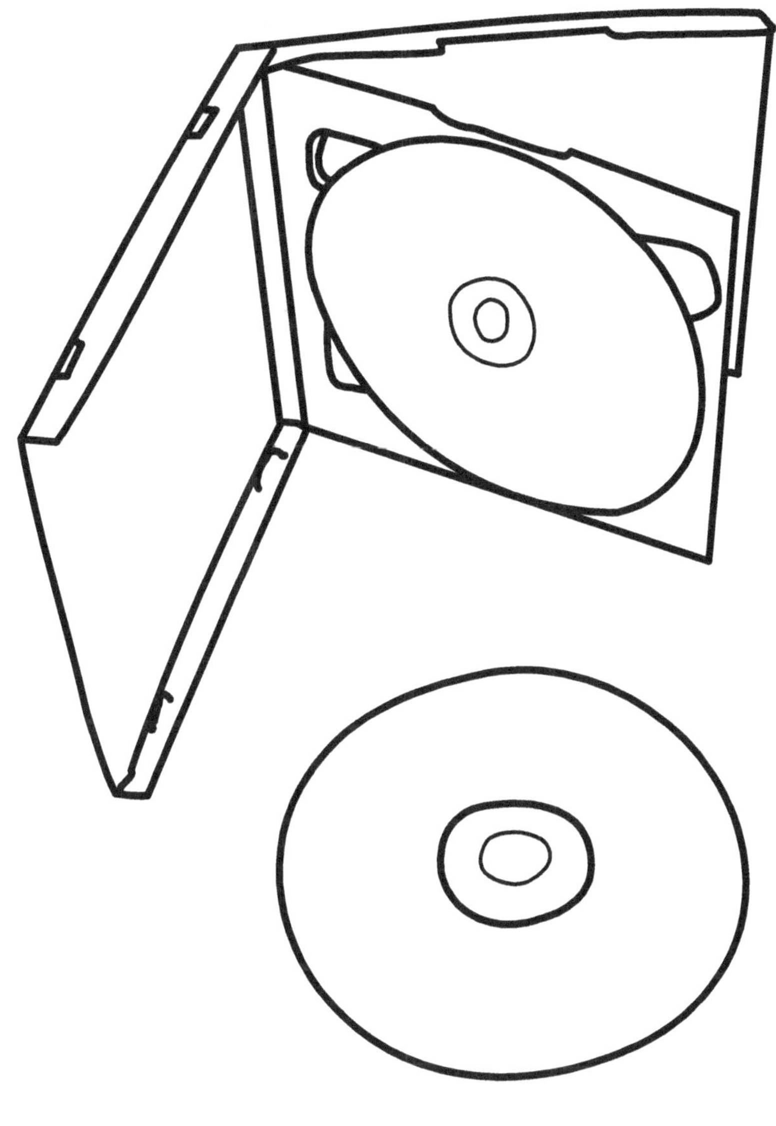

Fun fact
Mini CDs ranged between 2.4-3.1" in diameter and stored up to 24 minutes of audio

Fun fact
This format was later adapted to store data on the following formats: CD-ROM, CD-R, CD-RW, Video CD, Super Video CD, Photo CD, Picture CD, CD-Interactive and Enhanced Music CD

SONY
Discman

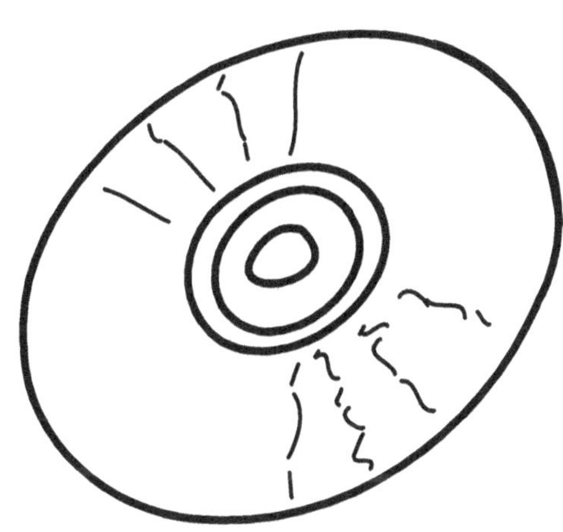

DAT

Format
digital

AKA
Digital Audio Tape

Size
7.3 × 5.4 × 1 cm

Era
1987–2005

Developed by
Sony

Capacity
180 minutes

Fun fact
This format looks similar to a Compact Cassette but around half the size

Fun fact
This format was lobbied against by the Recording Industry Association of America (RIAA) in an attempt to prevent high-quality copies

Fun fact
This format could record at sampling rates equal to, higher and lower than CD quality (44.1, 48, or 32 kHz)

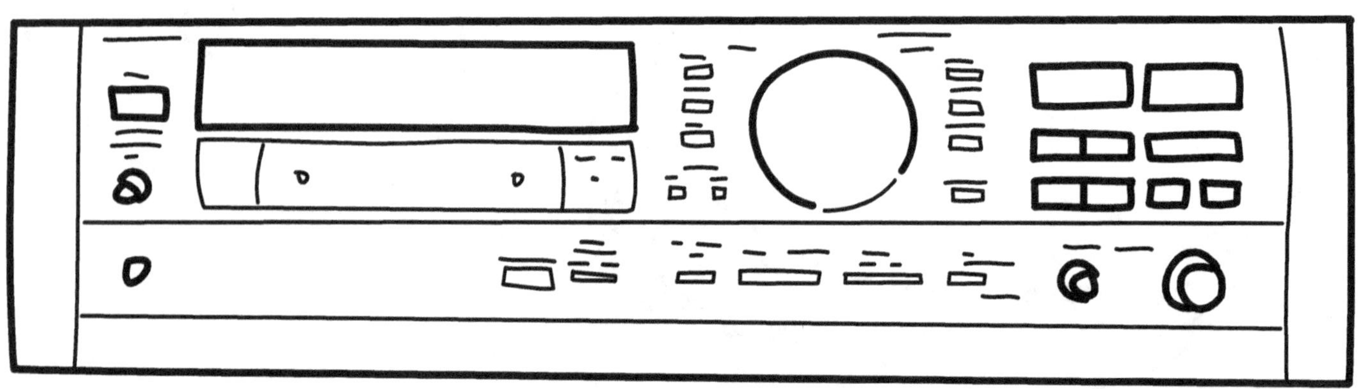

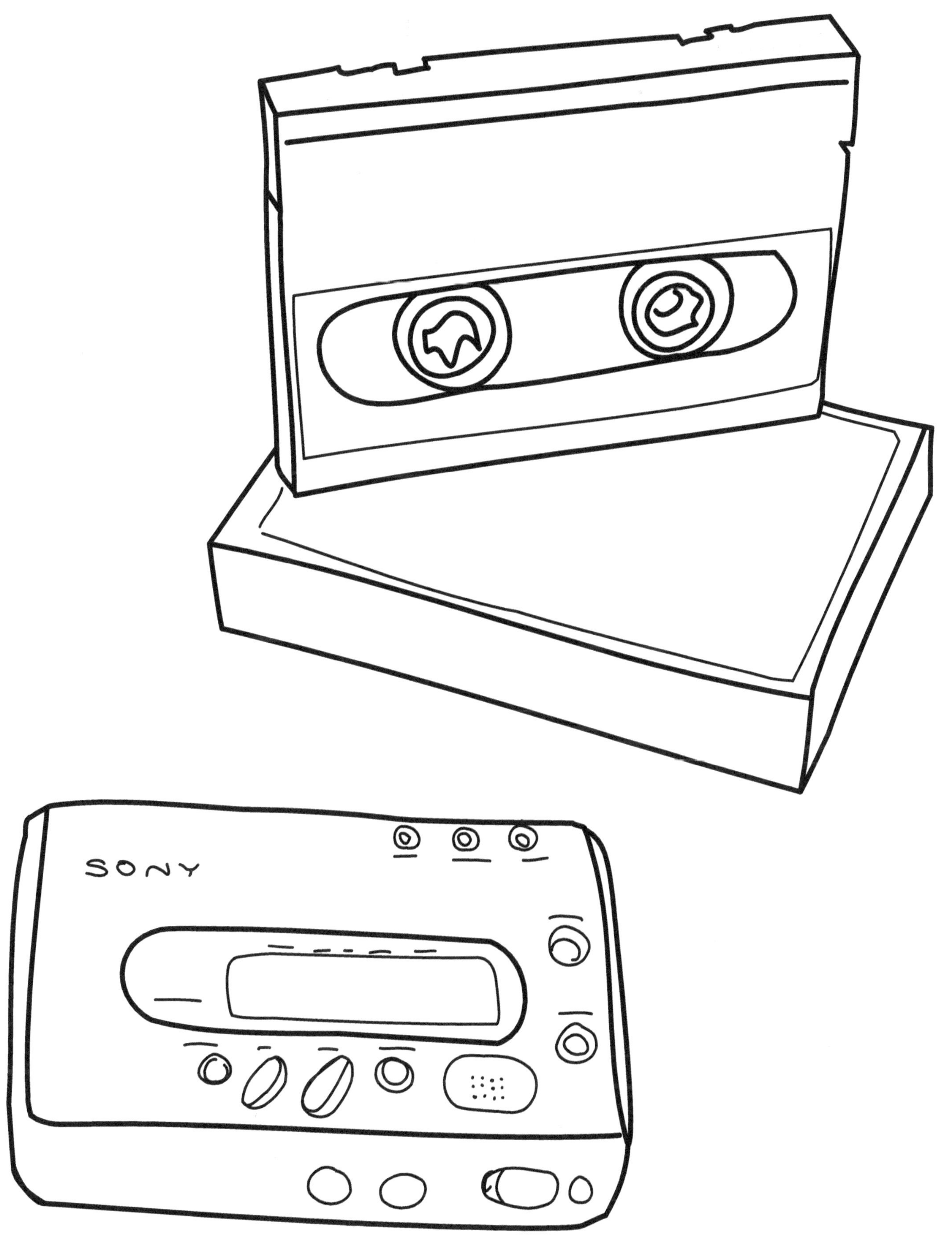

SONY

Pocket Rockers

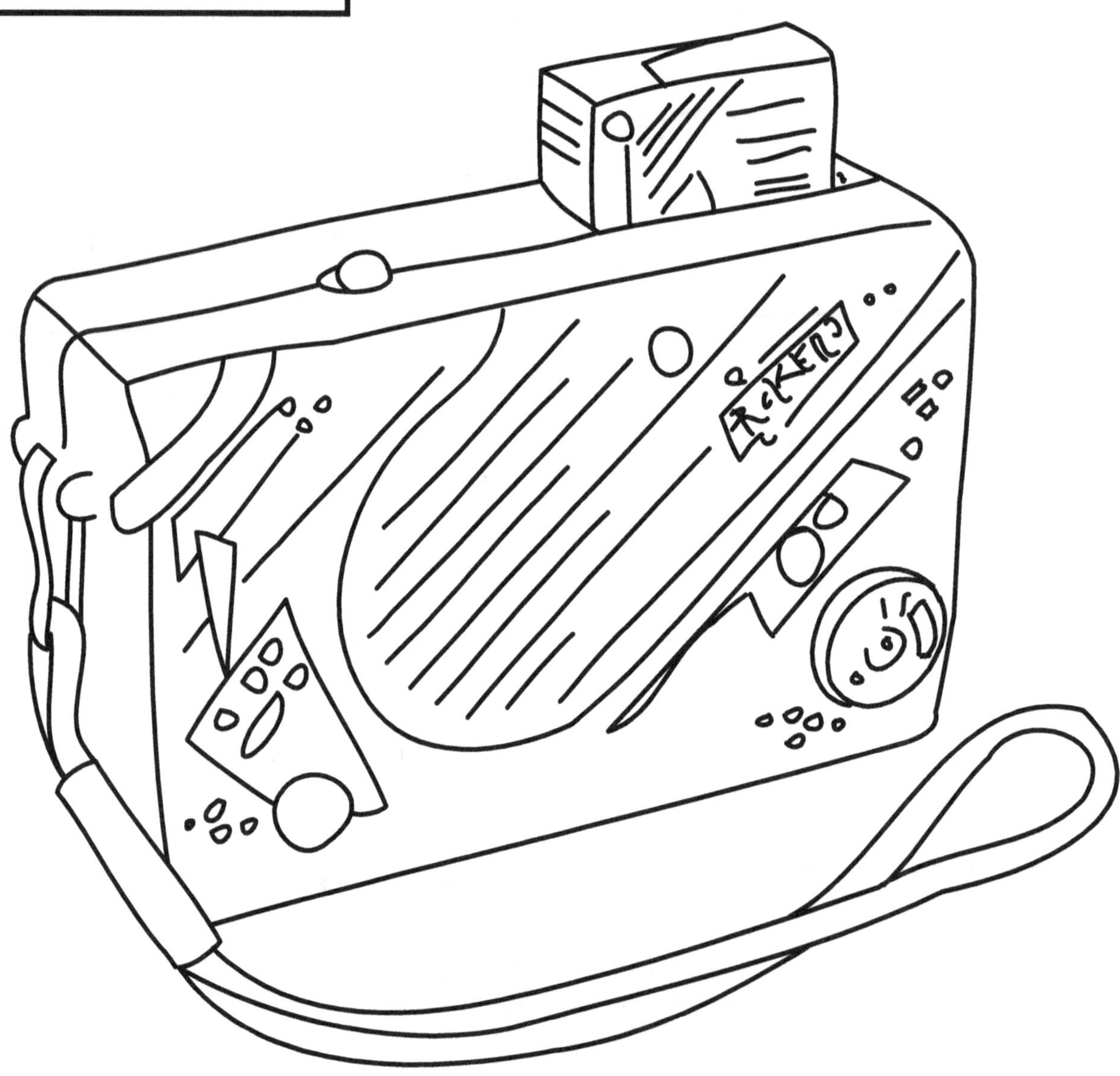

Fun fact
This was
a toy
format
marketed
to children

Fun fact
Each cassette
included
two songs
played
in mono

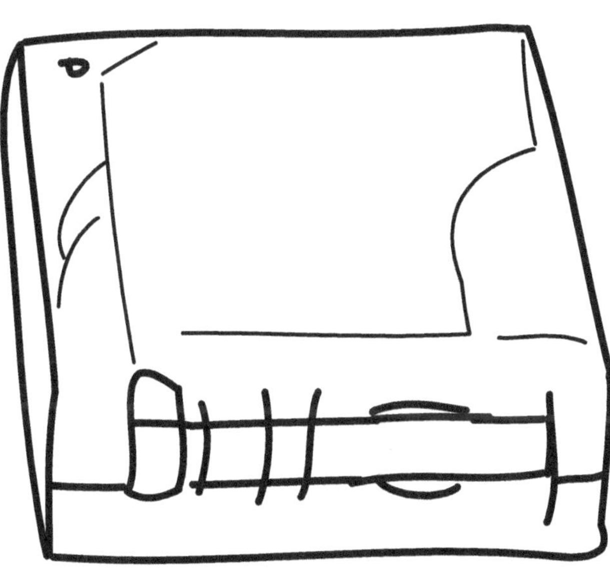

ADAT

Format
digital

Era
1992–2003

AKA
Alesis Digital Audio Tape

Developed by
Alesis

Size
18.7 × 10.2 × 2.5 cm

Fun fact
This format targeted the professional recording studio market

Capacity
40 minutes

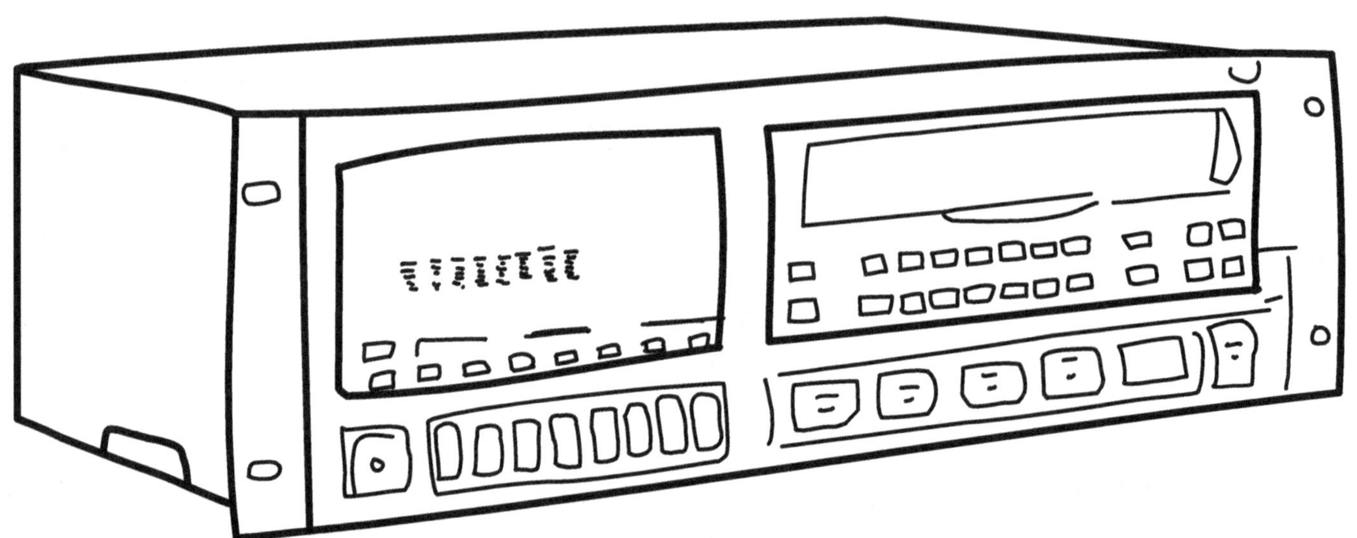

Fun fact
This format's
recorder used
S-VHS cassettes
as the recording
medium

Fun fact
This format could
support recording
up to 8 tracks,
but it was possible
to connect more
machines together
and create recordings
with up to 128 tracks

ADAT

DCC

Format
digital

Capacity
105 minutes

Developed by
**Philips &
Panasonic**

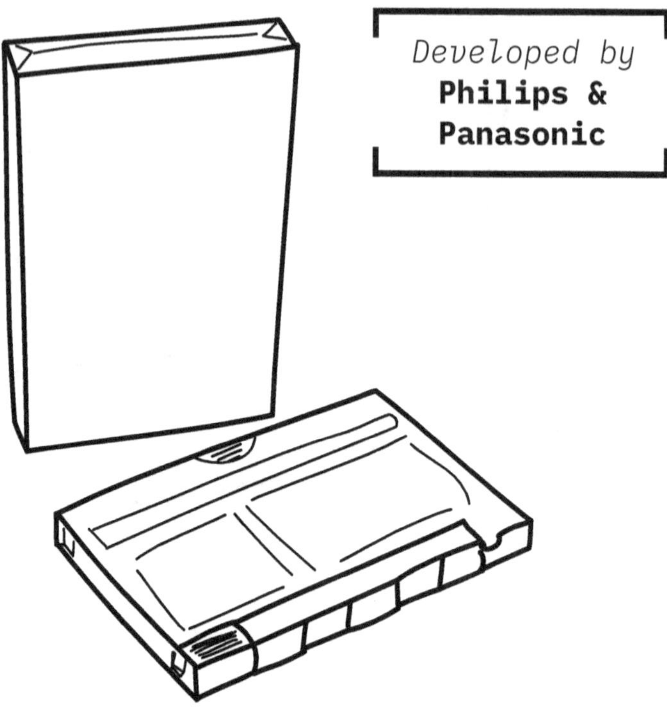

AKA
Digital Compact Cassette

Size
10.16 × 6.35 × 1.27 cm

Fun fact
**Each tape has
nine tracks per side
with eight tracks
for the audio
and one additional
track for auxiliary
information (track
metadata)**

Fun fact
**This format's players & recorders
were auto-reverse, meaning
every player had to be
able to position its heads
for the A-side as well as
the B-side of the tape**

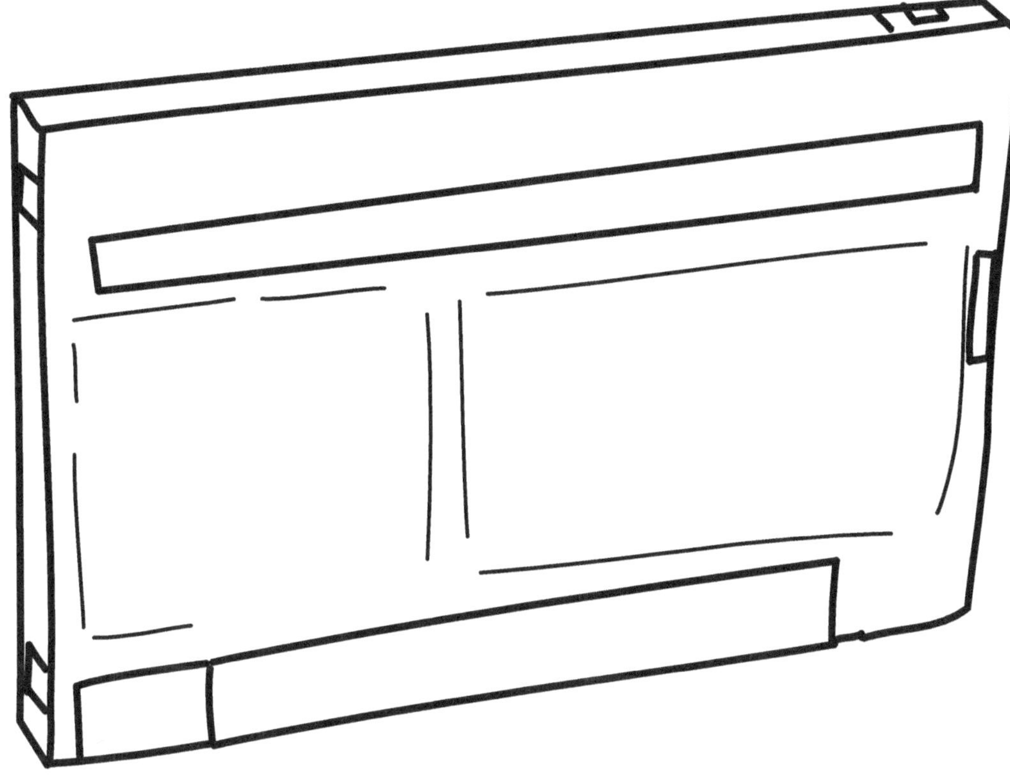

Fun fact
This format's shape was similar
to the analog Compact Cassette,
and could play back either type

MiniDisc

AKA
MD

Format
digital

Size
6.8 × 7 × 0.5 cm

Capacity
80 minutes

Developed by
Sony

Era
1992–2007

SONY
WALKMAN

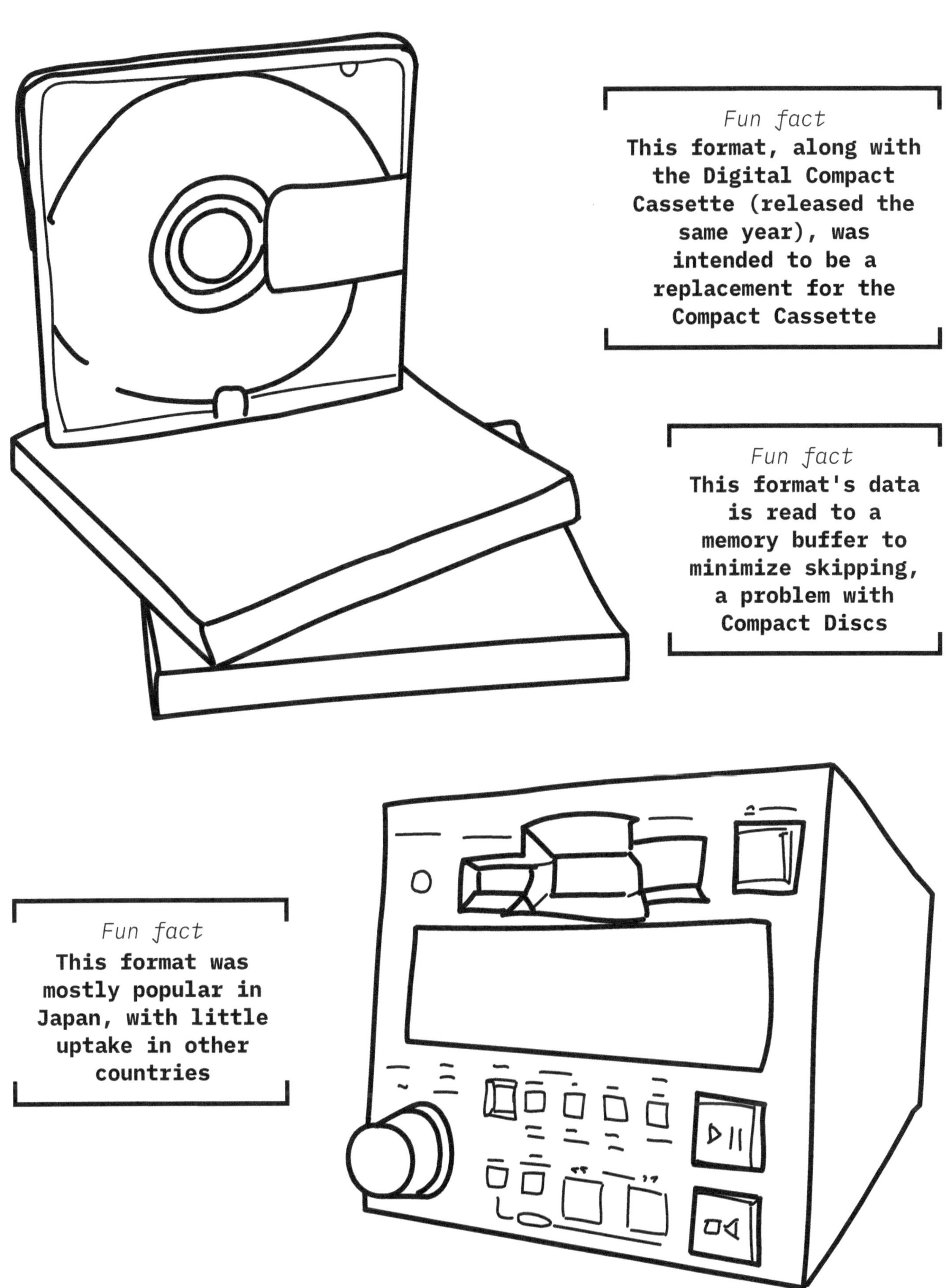

Fun fact
This format, along with the Digital Compact Cassette (released the same year), was intended to be a replacement for the Compact Cassette

Fun fact
This format's data is read to a memory buffer to minimize skipping, a problem with Compact Discs

Fun fact
This format was mostly popular in Japan, with little uptake in other countries

DTRS

Format
digital

Era
1993–2012

Developed by
TASCAM

Size
9.5 × 6.25 × 1.5 cm

Fun fact
Audio data stored in this format on Hi8 video cassettes allowed for up to 108 minutes of continuous recording per tape

AKA
DTRS, Digital Tape Recording System, DA-88

Capacity
108 minutes

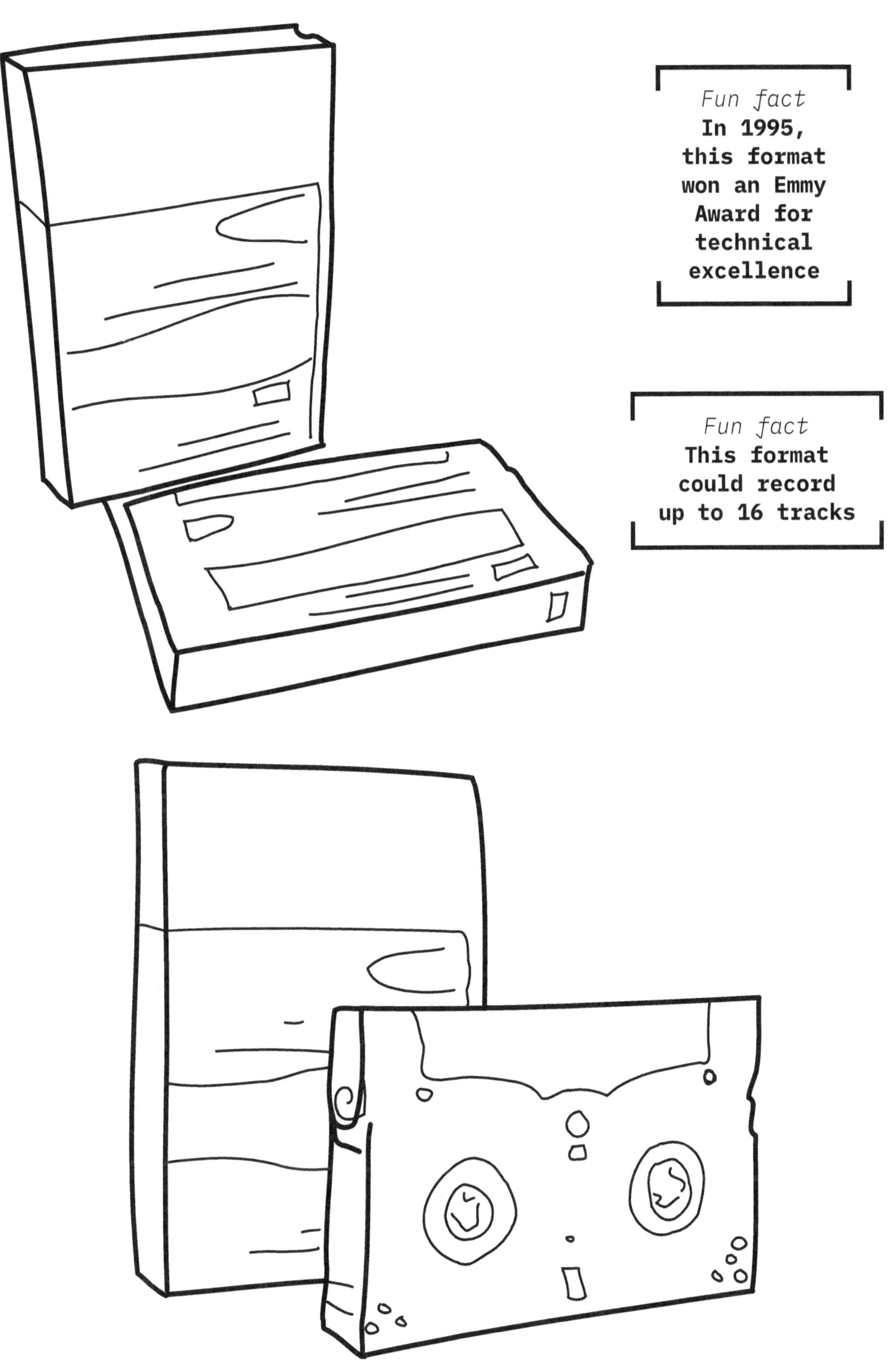

Fun fact
In 1995, this format won an Emmy Award for technical excellence
Fun fact
This format could record up to 16 tracks

Digital media player

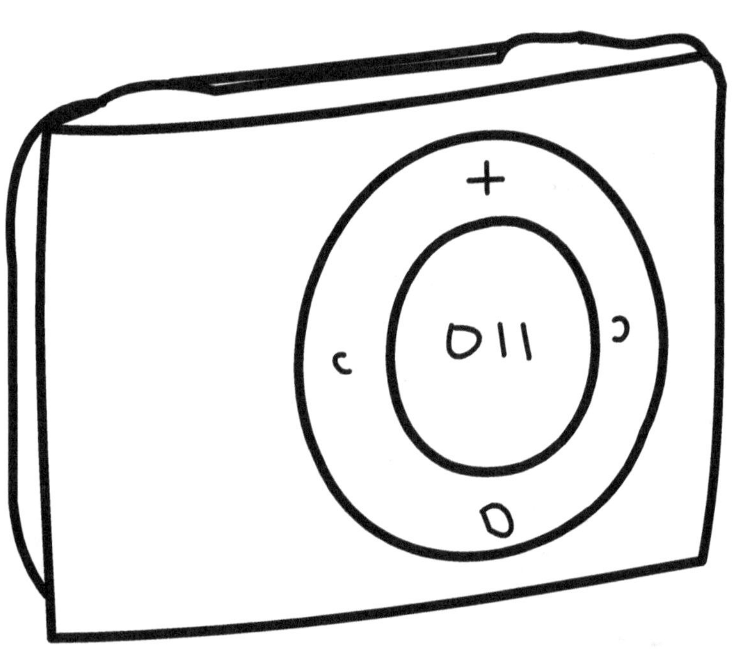

Fun fact
This industry was
largely defined
and popularized
by the Apple iPod

Fun fact
This format is marketed
as "MP3 players", but it
supports other
popular digital
audio formats
like WAV, Windows
Media Audio (WMA),
Advanced Audio Coding
(AAC), Vorbis,
FLAC, Speex and Ogg

Acknowledgements

Thank you to my technical reviewers, Andrew Weaver and
Susie Cummings, for lending your practical expertise and
camaraderie.

Again thank you to Rory: for endless listening to my hopes
and dreams, supporting every one of my ambitions, and for
everything.

Also available

The Illustrated Guide to Video Formats
The Illustrated Guide to Film Formats

About the Author

Ashley Blewer is an archivist,
educator, and software engineer with
over a decade of experience working in
video. Ashley specializes in video and
audio formats, digital preservation,
and communication.

Learn more at
ashleyblewer.com

www.ingramcontent.com/pod-product-compliance
Lightning Source LLC
Chambersburg PA
CBHW080331030726
47593CB00010B/2967